Modern Communications Systems

Modern Communications Systems

Third Edition

Research Triangle Park, NC

Lulu Press

ISBN 978-1-312-93472-6

Modern Communications Systems

For my family, Mimi, Donald, Mia, David, and Michaela. My parents and siblings have always been instrumental in helping me along life's path.

Cover Art : A picture from the NASA website of a ground station for satellites. The student of communications can learn a lot from NASA communications systems. They have mastered very long haul communications into deep space and are the best government source to look to for advanced communications systems revolutions. NASA first used CDROMs to record information from the moon. They transmit a lot of data through space on current probes and satellites and represent the highest echelon of human communications.

Modern Communications Systems

List of Figures

Modern Communications Systems

Modern Communications Systems

Modern Communications Systems

Modern Communications Systems

List of Tables

Modern Communications Systems

Table of Contents

Modern Communications Systems

Modern Communications Systems

Forward

I have found no book that adequately describes the technologies and
communications environments of the government or military except Dave
Adamy's EW 101, EW 102, EW 103, EW 104 series from AOC and Journal of
Electronic Defense. Crosstalk Magazine at Hill AFB Software Technology Center
is also useful. The service academies now teach network cybersecurity. This
book is an attempt to augment the books assigned to college level Data
Communications courses at local Maryland community colleges who teach
military and civilian students. It is not a primer on communications systems nor
the techniques used in most data communications and network management. It
does however, cover many of the topics often left out of other text books that
discuss the issues in computer systems that are managed and operated by the
federal government, military, and by private contractors. This book will benefit
them the most by explaining the regulatory environment and nature of agency
communications work from an integrated managerial approach. When we look at
the bottom line the technologies are becoming more closely connected to the
private sector as our suppliers. The unique requirements of the military make it a
separate entity when discussing military communications systems.

It is hoped that this book is beneficial and useful to the students of
communications systems from a managerial and technical viewpoint. It has
focused on a high level of how we manage our systems and not too many

technical details. In this regard it is not recommended as a primary text book for communications systems classes. Any mistakes are errors are mine alone and can be corrected by contacting me at dchiarella56@gmail.com. I hope you enjoy the book as much as I did writing it.

Special thanks go to my wife Mimi who has patiently listened to me ramble on about the works of my books and my kids who help me with living a graceful and meaningful life at home. She has let me take time that would otherwise be spent with the family to write these books and I am much appreciative of this fact. She shows courage of a rare type everytime she teaches her children in the sixth grade. I have often felt like one of her projects.

Modern Communications Systems

1. Imagine No Electronic Communication

Prehistoric Man

Prehistoric man had no means to communicate orally let alone electronically. Marty Barrack describes ancient prehistoric man's communication as a series of grunts and groans that was hardly discernable. He spoke to obtain help hunting and making weapons from fellow tribesmen. He did not have vocal cords capable of speech. Imagine if you may a world where no one talks a coherent form of language yet communicates with hand gestures and emotions. The Museum of Natural History has displays on this type of human on the plains of Africa and Asia where human life first began also called the "Cradle of Civilization". Scientific experts have recently said early man may have come from Eastern Africa. Moses is depicted as a man of northern African race by theologians. If the world truly began in the cradle of civilization or Mesopotamia, then the earliest cities had people who could barely communicate. Writing would have been a non-existent entity except for pictures on the cave walls. This would have represented real world events that early humans saw in their daily lives. Writing implements would have been sharp stones of various colors that would show up on the cave walls. This is hard to imagine by modern standards of cell phone communications, computers, and Cable TV. How far the human race has come!

Modern Communications Systems

Babylon –

Babylonia was one of the oldest societies in the world. It is documented in the bible. The Tower of Bable was created to give all mankind various tongues and unique identities. Ancient Kings like Nabechanezzer ruled Babylon with an iron fist. There were many Jewish people who lived in ancient Babylon. The Euphrates and Tigres rivers were home to what is believed to be the first city in Babylon – modern day Bahgdad.

The Greeks

The Greeks were astute scientists, mathematicians, and warriors. Atheians were considered academicians where as Spartans were considered the greatest soldiers. Men from each major city had a different attribute. All the tribes of Greek came together under the reign of Agamemnon. The ranks of the Greek army communicated with a horn and by voice commands that were not to be disobeyed (with a penalty of death). The transported messages by chariot. During the Trojan war, the Greeks undoubatbly used navigation by the stars to find Troy by sea. The original Greek word "alphabet" starts with the words for the first two letters "alpha" and "beta". Greek symbols are used today in mathematics and give us a rich mathematical language spoken by many regardless of culture of origin. The very fact that Troy's Aneas founded Rome indicates the tight link between the ancient world of Greece and the Romans.

They also shared the same concept of Multi Mythological God religious beliefs only the Romans used different names than the Greeks. Jupiter in Rome was Zeus in Greece, King of the Gods. Mercury was messenger of the Gods. The Romans spoke Latin which used some Greek words and then was transformed into today's English language that we use with multiple cultural roots in many previous languages. Ancient Greeks valued education and sciences. They passed this love onto future generations and societies. Archimedes, Plato, Aristotle, and Socrates were all great men of sciences and philosophy. Achilles, Ajax, Odysseus, Menelaus, Agamemnon were all great warriors and Kings of the Greeks. In deed, they are documented in The Illiad, The Odyssey, and The Aniad by Homer and Virgil, the two most famous Greek poets of the day. Today Greek translations of the Holy Bible are very useful to our Christian scholars in helping us to better understand the ancient world and it's changing beliefs, people, and cultures. The Greek Orthodox church is a very strong sect of the Christian church global system.

Egypt –

Egypt played a major role in ancient times as the major country in northern Africa with oral and written traditions. Egyption hyroglyphics were written on the tombs of the Pharoes and the Pyramids at Giza. It is now believed by the History Channel that only loyal Egyptions were entrusted, not slaves, to build the pyramids. This does not account for Jewish folklore. It was considered duty to the King and done with all the reverance of today's military service in any of the

civilized countries. The reward was life ever lasting for the builders as well as the

king entombed forever. Egypt has the only remaining wonder of the ancient

world in the Great Pyramid at Giza. It remains as the most perfect reminder of

engineering capabilities of the ancients. This included all types of

communications required to plan, and execute the building project itself. It is

thought that papryrus is an egyption form of writing. Ciphering and encryption of

important messages was also done in Ancient Egypt during the time of Cleopatra

and Mark Anthony.

Old Testament Times

Much of the old testament occurred from the Jewish Year 1 until the

present day Jewish year 6757. The books of Moses describes stone tablets with

the laws of Moses scribed by the hand of God. Scientists have discovered that

the tablets may have been stored at the base of the real Mt Sinai in a History

Channel Special. The Air Force was said to have discovered Noah's Ark on Mt.

Ararat halfway up the mount. Genesis, Exodus, Leviticus, Deuteronomy, and

Numbers form the Pentateuch or five books of Moses. It is my belief that science

can verify many of the historical events of the bible through modern archeology.

The bible is one of the single most read books in the modern age and as such

should be considered one of the greatest books ever written. Books of other

religious traditions are also important to other cultures of people as historical and traditional communications that allow passing along of information from one generation to the next. The early Jewish oral tradition of telling the stories of the bible was all that was available in the early days before Christ and his disciples. The Torah is the written laws of Moses that is read from in today's Jewish ceremonies. The Koran is the Muslim world's bible and is worth reading to understand that extremists prevalent today are not of the mainline belief system.

Jesus Christ's Time

What we know about Jesus' time is captured in the new testament of the Bible which was written about 40-60 AD. This means there was a time lag in the writing of the Gospels themselves by the original authors (Matthew, Mark, Luke, and John). Jesus' language was Aramaic and had to be translated into several languages (Greek, English) to be in the form we study today. There is great speculation that a lot of what Jesus actually said in his teaching parables was not included in the bible and that the disciples may have collaborated on their stories about Jesus and his miracles. It is clear that writing on parchment or papyrus was a new standard for peoples of this time frame. The jar found in the desert that represents the lost books of the bible known as the Gnostic books are significant to the religious world as they may hold secrets about what Thomas (the Twin) and others close to Jesus said and experienced with him. These books of the bible were not allowed to be placed in the bible by church authorities and thus became secret communications of the early church after they were banished as heretical.

Roman Empire

The Roman Empire relied on paper messages called papyrus. Every so often the emperor would get an encrypted message from another country who had the means to create such a message. The emperor had his special seal he put on messages to ensure the reader knew it was him. David Kahn discusses some of the ancient cryptography devices in his 1968 book "The History of Secret Writings" which also discusses the NSA. Indeed the Dead Sea Scrolls are written on papyrus and preserved as well as possible for a paper product. It was an improvement over other methods of the time which included writing on cloth (The Cloth of Turin) which had an imprint of Christ's face upon it and other cloth writing. Accounting recording had made in-roads during the Roman empire also and this was also done on papyrus. Early Christians became good as scribes and realized the importance of preserving history of what they had witnessed. These early writing were secret and protected through the centuries by notable such as the Knights Templar, The Vatican, and Western Protestant Church authorities into the document we know today as the bible. The Roman soldiers in the legion wrote home frequently from outposts in the empire. They were the only ones paid a steady salary and thus had the means to communicate with family far off. The letters were delivered via a crude Pony Express system.

Modern Communications Systems

Ships, Roads

Ships and roads developed as civilization expanded from ancient Rome to the west. The Romans were significant because they systematized travel and communications along these travel routes to the provinces. As Rome conquered other countries they provided for travel and communications. The oldest highway in Rome was the Appian Way. Rome had great warships also which allowed it to travel the world on the seas. Italy as a peninsula is strategically placed in the Mediterranean. Roman highway and ships also meant Roman engineering. Rome spread all the way to Britain and established the city of London as one of it's early outposts[1] in 43 AD. The Romans built Hadrian's wall to the north of England as a divider to Scotland which Rome never did conquer. The Roman's brought architecture and engineering methods to Britain including early communications techniques by messenger on highways, bridges, and rivers. The Thames river was strategic as a defense and travel route through London. The structures in early London were built with stones from the quarries on the Thames river. Roman Mark Anthony sent many encrypted messages to Cleopatra Queen of Egypt when he fell in love with her. The Egyptians themselves were experts at encryptions in secret communications. There were also a very script based society and kept records on the stone walls on crypts and tombs for today's archeologists to decipher.

---

[1] Discovery Channel Special on London

Modern Communications Systems

Medieval Times

During medieval times messages were still sent by horseback rider and usually coded or entrusted to a person who knew the sender or maybe was servant of the person sending the message. Kings and queens would send only the most trusted of servants on these missions. When a receiver did not like the message they may often strike out at the messenger. From this came the term "Do Not Kill the Messenger" if you do not like the message. The medieval way of communicating involved much formality as few people knew how to write and read. Music was well developed by those who learned how to play it and oral traditions were carried on by families in feudal society. Messages were carried by personal courier on horseback and often they were encrypted by some cipher or special invisible ink.

17th century Scottish Enlightenment

The 17th century Scottish enlightenment period brought many good things to the world including new religious beliefs in freedom, new architecture methods, trade to the new world colonies (America), new engineering methods, new mathematics, and new inventions to the world. Scotland was a poor country who had a great affect by exporting democracy to the world. Colleges became better developed with new ideas during this timeframe throughout Europe. Other countries, especially the United States, took the best ideas of the people during this time period and made it better. In Herman Author's book, "How the Scots Invented the World", the author tells the world all about this period in time when famous Scots affected the world. Some of the most famous Scots in America were Alexander Graham Bell, Andrew Carnegie, Robert Fulton, 17 U.S. presidents, and many more. The basic ideas on freedom in the world come from the Scottish fight for freedom in 1296-1314 against the English domination. This included the struggles of William Wallace and Robert The Bruce. However, as any student of Scotland knows there were many kings who gained their power by killing their predecessor. Thus the Scottish royalty highly enforced religious beliefs among the common people to ensure a peaceful internal nation. This too was exported to the world. Common sayings were exported to the world. Everything from sayings like "Be yourself" and Robert Burns poems to songs like "Scotland The Brave" and "Amazing Grace" were exported to the world cultures. These had all had a great impact on improving human and technical communications in the world of that time period.

The American Pony Express

The American West movement saw the rise of the Pony Express and stage coaches for deliver of passengers and mail from the East Coast cities to the West Coast new cities. The mail was protected by a stage coach driver and a man riding with a shotgun. US gold reserves and bank money were also shipped by stage coach which gave the shotgun man even more to protect against the possible robbers along the dusty way. There was a fee sending the mail this way. The Wells Fargo company managed the service as an early contractor of mail carrying to the west. This was until the telegraph was built. The telegraph augmented communications but did not replace the written mail letters. It was used to keep trains on schedule. Trains also carried the mail to the locations in the west. They would have large safes in the mail car that would be protected by a Pinkerton detective man. The telephone eclipsed usefulness of the telegraph but people learned they liked being able to wire someone money and this service is still available today in banks and post offices. During this time in America, communications and transportation had parallel developments that helped American society and the world, in general. The Pony Express idea may have originated with the Circuit Riders in the colonies who rode from town to town preaching the word of the Lord. Paul Revere was one of these Methodist preachers who kept the colonies in touch with the greater world at the time before he made is famous ride into history exclaiming "The British are Coming!".

US Postal Service

The original American post office was setup by Benjamin Franklin in Philadelphia. It was just one of the many great things he accomplished in his lifetime. He was one of the greatest non presidents America ever had. He had many talents and wore many hats. His early life experience as a printer in his Boston family upbringing gave him experience with printed press. One of his brothers was a printer. He built his own printing press for use in Philadelphia. He reviewed publications by others in the colonies. Later the Pony Express was established for communicating with the west. In the 20th century, the postal mail began to be carried by automobile and airplane in 1922. This evolved into the great postal system we enjoy today. Many private companies are available to augment the postal service's capabilities. The future may hold US Postal email and internet email. One can already purchase a roll of US stamps over the internet postal website (http://www.usps.com) and have them delivered to his home. The postal service would be wise to help build postal services on internet and Dr. Franklin would be proud of the use of his invention of electricity!

20th century and the Industrial Revolution

The 20th century brought about the industrial revolution in major countries in the world. New ideas to automate society contributed to new inventions that helped fuel the democracies. The cotton gin, light bulb, telephone, telegraph, steam engine, gas combustion engine, Gattling gun, and many other inventions revolutionized the world. Things were done before were made easy with these inventions. All nations contributed inventors who made the new world inventions that industrialized many industries such as military, transportation, farming, banking, and many others. Faraday and Maxwell created theories on electromagnetism fields. The focus was on improving mankind's condition in the world through creating new machines. In the 20th century this continued with the new inventions such as aircraft, trains, automobiles, electricity, atomic energy. World class physicists help develop new systems that have helped mankind and been used for military scientific purposes.

The Computer Age - Second Industrial Revolution

The age of electronic computers from 1940's through 2000 were dubbed the second industrial revolution. The non-electronic computer dates back many years before the 1900's to the Abacus counting beads on a stick. The Japanese used something called a Sordoban which is similar 5 rows of counting beads and still in use today. The machines of yesteryear needed to be controlled by computers and computer software and the war had to be won quickly. People became programmers and computer scientists who understood the first industrial revolution and how to get computers to think and drive the machines we use in society. Now computers are so commonplace we all expect them in every phase of life to help us with our operations and control of almost everything. It's hard to find anything that can not be computerized. No computer is as complex as the human brain yet. They focus on fast calculations that tire the normal human mind. Large parallel computers process arrays that could not be done by humans. Humans will always control computers and for this reason those who used to be factory workers needed to retool and learn more about programming and systems analysis and building the computer brains of the second industrial revolution. Computers and computer chips became the driver for many types of

Modern Communications Systems

communications equipment including military and commercial applications. World wars caused the advance of communications computers such as Colossas. Radar was also invented during World War II. Embedded computers and computer and comunications software do much thinking in today's mechanical systems including landing aircraft, driving robotic cars, drones, and pilotless vehicles. Anything that can be communicated can be computerized, encrypted, and sent as a message. It's a wonder what America will be like in 2050.

Modern Communications Systems

2. Communications and Systems Theories

This chapter focuses on some of the theoretical concerns of systems and communications. The student will hopefully understand the types of theories that have led to the production and management of major communications systems.

General Systems Theory – GST

The father of General Systems Theory was Von Bertalanfyy. The idea was that systems thinking could be applied to any science. Think of the systems we have developed through sciences – the solar system, the ecological system, biological systems, anthropological systems, evolutionary systems, anatomical systems, medical systems, weapons systems, transportation systems, food systems to name a few types. Aristotle said that the "sum of the parts was greater than the whole". In systems sciences we break things down into parts and improve them. The 5 basic premises of GST are listed in figure 1.

Figure 1. The 5 Basic Premises of General Systems Theory

P1. Order, regularity, and non-randomness are preferred to lack of order
or irregularity (chaos).

P2. Orderliness in the empirical world is good

P3. There is order in the orderliness to the second degree (i.e. laws about
laws)

P4. To establish order, quantification and mathematics are valuable
aides.

P5. The search for order and law necessarily involves the quest for the
empirical referents of this order and law.

The systems approach is therefore a scientific way of looking at the world
using orderliness, natural laws, and mathematics.

Information Theory

Information Theory is a theory that delineates between data and
information and explains how data is transformed into information to give us
more information and meaning. This is the premise of every MIS course in
college. Norbert Wiener was quoted as saying "Information is the name for the
content of what is exchanged with the outer world as we adjust to it, and make

our adjustment felt upon it". Data is transformed into information that a decision-maker can use and then actions are taken[2].

As the cost of information increases the utility of the information decreases. That is, when we look at the marginal value of information we see that it must be fresh and concise to help in decision making. The utility of information to impact decisions in organizations and governments decreases as there is more information.

We learn early in our systems careers that you have various information flows for decision making – daily, weekly, monthly, bi-weekly, quarterly, semi-annually, annually, bi-annually, tri-annually, and long term projections of 5 year plans, 10 year plans, 20 year plans, etc. Interestingly enough these long term plans are based on long term historical trends data going back at least the same among of years as one is projecting into the future. There simply is no other way to predict without looking at trends analysis using the mathematical tools of economics, statistics, and probabilities to aide in decision-making.

Communications Theory

Communications Theory tells us that if we have more nodes, we have more complexity and we may have more errors. It's like the message in the classroom that the teacher gives to one student on one side of the room and then

---

[2] Schoderbeck & Schoderbeck, Management Systems: Conceptual Considerations, pp. 142-3

every student passes it along to the next. When the last student tells the

message it may or may not be garbled when it is received after going through so

many senders and receivers with different perceptions and filters (figure 2).

Figure 2. Basic Communications Model

Sender --------------------------------→ Receiver
                Filter and Channel

Figure 3 shows the 6 node network that illustrates the complexity as the

number of nodes increases.

Figure 3. 6 Multi-node Network

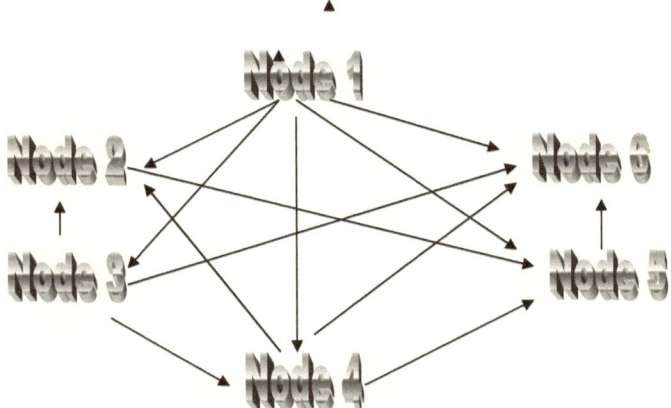

Modern Communications Systems

The formula that determines the number of connections is called the Graicunas formula and is represented in figure 4. It says that the numbers of potential connections paths, C, can be determined by this formula for any given number n of nodes.

Figure 4. Graicunus formula

$$C = n [2^n / 2 + n - 1] \quad \text{or} \quad n [2^{n-1} + n - 1]$$

For example,

Where $n = 22$ $C = 46,137,806$

$n = 300$ $C = 3.05 * 10^{92}$

$n = 3800$ $C =$ calculator error

It is easy to see how complex a communication system can get very complex very fast. This applies to people talking to each other as well as data systems. The larger a company the more complex the communications becomes and the harder it is to have a direct relationship with every person in the company. This has implications for leadership styles and the need for electronic communications systems that traverse time and space to let people talk asynchronously.

Modern Communications Systems

Organizational Theory

Organizational Theory is the study of human organizations and how they affect work processes. It is the contention of this book that communications as one of the seven classical managerial functions is a vital component of organizations. Surely the organizations who have electronic communications are far superior to those without electronic communications. Mintzberg employed 5 different types of organizations. They were:

Figure 5. Types of Organization Structures

1) The Simple Structure

2) The Machine Bureaucracy

3) The Professional Bureaucracy

4) Divisionalized Form

5) The Adhocracy

Frederick Taylor was the leader of the classical organizational theory. Mayo was the leader of the Neoclassical school. The Modern Systems Approach was pioneered by C. Barnard and Max Weber.

When an organization reorganizes it is a sign that organizational changes are desired by upper management through the means of realignment of human resources. Not every reorganization is effective and surely after too many reorganizations in a short period of time the staff and workers think much less of the management's ability to effectively lead. It is true that graduate schools are teaching reorganization as an effective tool for managers. It is also true that when a reorganization occurs, there are winners and losers from the older organization. This is what many people remember when they adapt to the new organizations implemented by managers. It is important to remember that teams can cut across organizations and that sub organizations develop to meet larger goals. Committees can be effective in meeting organizational goals in any of the structures in modern organizations. An effective email systems can provide the systematic communications that can enable virtual committees and virtual knowledge workers contributing across the miles to a committee goal.

Guidelines for a Reorganization

When management decides to reorganize a group of people, the following steps are suggested to ensure successful implementation.

1) Study the organizational problems and where information is not flowing properly between groups. Study the past information flows and decision points in major team leaderships. Make no reorganization before you have completely understood the current organization and how it functions or should function

smoothly. A typical time frame is after six months on the new job where the organization needs restructuring.

2) Bring all the latest technological elements to bear on the reorganization problem by including subgroups who have new information and knowledge to offer the new structure. Borrow structures from other groups that have reorganized into more efficient groupings. Map out the type of organizational structure that will fit the goals and objectives of the group being reorganized.

3) Obtain top management approvals to do the reorganization. It is very difficult to reorganize without top management approvals. Likewise, discuss with subordinates past structures that have worked and not worked as well as products that have been well accepted and those that have not.

4) When reorganizing create three alternatives and proposals for the new form of organization and test them against senior staff and organizational goals and objectives. Ensure resources are available to complete each one of the alternatives. Show the alternatives to top managers who approve the process to let them know you have considered alternatives and which plan you think will work best. Make sure you document each step of the way using a visual diagramming tool of some sort if you have computers.

5) Get buy-in from staff that this is the right thing to do at this time for the organization. If they are a part of creating the new structure they will conform to it more readily. Understand that people need time to get used to new ideas and they want to operate more efficiently, but may not know how the big picture fits together or how they fit into it.

Cybernetics

Cybernetics is the field of is concerned with control and feedback to the manager or leader about his organization or processes under his command. The military sometimes calls this command and control. Computers are more often providing automated command and control in many situations in the technical world. Human intervention is always needed as a check on the computer processes. When a decision is made the outcomes are checked for accuracy and compliance with the intentions. If there is variation, then a new decision needs to be made to adjust for the variation. Figure 5 shows a first order feedback control loop as just described.

Figure 6. First Order Feedback Loop

```
Goal ----→ Summation Point --------------------→ Process --------→ Outcome
                  ^                                  |
                  |                                  |
                  |_____|
```

Feedback Loop

Modern Communications Systems

Since systems can make basic decisions and decisions can be programmed onto computers, these first order feedback systems may be automated or non- automated. Sometimes the real time nature of the feedback requires the computer to make the adjustments pre-authorized. Other time the adjustments are made in the decision making process by human intervention without pre-programmed decisions based on environmental sensors. A chain of command in an organization gives the authority as to which level of automated decision making will be allowed. This will usually involve electronic systems of many varying types designed to meet the organization's goals and objectives.

Key to the concept is knowing what results one manager would like to get from the process. If one has a good idea of what he is looking for, then he can work towards achieving that end result or outcome from the process using iterative feedback looping with people and machines. Cybernetics and control feedback loops is the origin of the saying "Keep me in the loop" which refers to the managerial communications loop in this case. Note that various managers may or may not choose to keep others in the loop based on what outcomes are desired and what the organization structure and goals are to be completed by the process.

Modern Communications Systems

Maslow's Hiearchy of Needs

Social scientists and psychology have given us many tools to better manage our technology people. Abraham Maslow's hierarchy of needs is a set of needs that every person perform under that must be met in order to advance to the next level and more fulfilling work. Maslow was a Jungian phychologist educated at the University of Wisconsin and wanted to create a model of normative behavior that explained the world as he understood it.

Figure 7. Maslow's Hierarchy of Needs

1. Physiological needs

2. Safety needs

3. Belonging needs

4. Esteem needs

5. Self Actualization needs

Once each of the lower needs are met, a person seeks the next higher level of needs. In communications systems, we are self actualizing by communicating for our advantange and gain of the organization we represent. The managers of a communications organization should therefore be enabling people to communicate and fulfill self actualization levels of human performance.

Modern Communications Systems

This is the only humane way to run an organization and help people enjoy their work and fulfill their God given potentials.

Modern Communications Systems

3. Mass Communications

Mass communications is basically communicating to the large groups of people in society via electronic media of the airwaves. Today this includes cable mass media and not just radio frequencies. Many colleges have courses that lead students to mass media jobs on TV or radio. One of the reasons President Reagan received the nickname "The Great Communicator" was because he was once a radio announcer and understood the potential for reaching millions of people's ears this way. These types of jobs shape the roles of viewers and listeners and are usually politically oriented towards the owners of the radio station or TV station and their agendas. Good unbiased reporting of news is difficult given this reality. Printed newspapers would also be in this category.

The Telephone

Alexander Graham Bell invented the telephone when he made that famous statement "Watson come here I want you". AT&T and Bell telephone won that first contract from the US government and the rest is history. The loser Gray communications never fully recovered from the loss of the original patent on the telephone. In 1934, AT&T started wiring the country at congress's request. In 1984, the telcommunications companies including AT&T were asked to split into smaller regional companies and avoid an anti-trust suit as a monopoly. Today, we enjoy one of the best markets in the world for telecommunications via telephone and this service is used as a backbone for basic digital communications on twisted pair wires which is often called voice and data grade lines. Optical wiring is being done of many counties and states to improve the quality of the system. Cellular systems are also one of the inventions that changed the way the world talks on the phone.

TV

TV is one of the great mass communications machines of the 20th century. TV has captured the hearts of Americans and the free world. We can see live what we once saw only in person or heard on the radio. Politicians can use TV like they never have before. President Bush studied TV and the President to learn how to run a campaign on TV while he was at Yale business school. This included lessons on the first TV election date between Kennedy and Nixon. Nixon lost primarily because he did not appear wise on TV and was rather uncomfortable. Kennedy on the other hand was very relaxed and came across confident and focused in the first TV debate. Satellite TV is used by colleges as an advanced means of delivering college courses to distributed locations from main campuses. Today, the best shows on TV are the educational channels and government channels which are usually on cable TV. TV comes in many frequencies and varieties in 2004 including cable, satellite, UHF, VHF, and closed circuit systems based on what you can afford to pay for.

Modern Communications Systems

Radio

      Radio was Marconi's invention but Hitler turned it into the political tool of mass communications during pre-World War II. Hitler, whom believed in the occult it is thought now, found a way to reach many people beyond the crowds who came to hear him. He brainwashed a country into a terrible war and mesmerized Germans on all issues of the day. His Nazi party gained control of the country and started the worst war in history for world domination. He took that country out of depression in the 1930's and started the Spanish civil war of 1936. Why did America let this happen and not be able to foreshadow WWII? We had no formal mass media reporters there to capture the injustices as we do today. Indeed radio was one way to capture early politicians. FDR gave his weekly fireside chats on radio on Saturday mornings, a tradition which every president uses to this day. Radio has evolved into satellite radio (XM and Sirius which are good investments). Packet radio is an invention of World War II. AM, FM, and shortwave radio are all methods of broadcast media we know as radio. Ham radio technology relates to the private communications by hobbyists.

Internet

      Internet has become the new mass communication method of the post 2nd industrial revolution. We are so good at mass communications that we now link computers and communications and talk on computers like a TV and radio combined in one. It provides a visual and audio rich environment for end users. Teleconferencing has been enabled by internet. Internet 2 is a project for satellite internet communications. This is also called the "Internet in the Sky" project which is due to be completed by 2005. Normal people and professors are learning HTML and building web pages to educate others and have an Internet presence on the web for advertising and marketing purposes. Internet has become the greatest tool on the new millennium for normal people. Research can be done on internet. Any topic can be researched. Children and adults love internet, but it must be regulated to keep the bad guys in check.

Internet is the output of a US Government DARPA project in 1969 released to the public along with other networks through academia. BITNET was another network for academia only. Today creating a web page is as normal for a business as any advertising activity in the news papers. Electronic newspapers can be viewed online immediately for free. Congress has discussed the taxation of internet but so far it has not been taxed due to a moritorium.

Modern Communications Systems

4. Long Haul Telecommunications

Long haul telecommunications is revolutionized by the satellite industry. The networks of satellites are crowding the sky. The definition of long haul telecommunications is cross border telecommunications between countries and political boundaries which may sometime include language barriers. The Federal Communications Commission (FCC) regulates telecommunications industries in America along with the Federal Acquisition Regulations (FAR, Title 48 CFR) which is regulated by the General Services Administration (GSA). Individual manufacturers comments are included in the regulatory process as well as directors of the divisions in the policy making organizations. The competitive companies are required to compete for business within the government on major IT Telecommunication projects such as FTS 2000, the government's telecommunications system. The technical and legal aspects of bidding on these contracts is the domain of the GSA and GSA Board of Contract Appeals (GSABCA). The companies have the technical knowledge and equipment to run the networks. Universities prepare graduates for both management and technical telecomm positions. Major companies can lobby congressmen for votes on positions on telecommunications issues. Congress holds hearings on the improvements in the systems they will fund such as the recent Voice/IP issue. These can be heard on CSPAN radio and are quite interesting to the layman. I am never amazed at how little the congressmen actually do know and I think they have so much on their agendas that any details get lost except at hearings, from sworn testimony, and through the Administrative agencies appointees. There was only one congressman in the 1999 CSPAN directory who was a computer programmer and scientist in his career and assume for the most part he would be one of the few qualified to understand the problems with telecommunications systems operations, network management, .and systems engineering from direct experience. The Science, Technology, and Transportation committee is responsible for making new laws in telecommunications. The recent chairman was Senator John McCain. The 1984 Telecommunications Deregulation Act let companies compete for long haul telecommunications and created the Baby Bells. AT&T had a monopoly up until that point in American history of telecommunications since the 1934 Telecommunications Act gave them authority to wire America for telephones. This telephone system is still in use today with multiplexed signals on the wire. Newer technology has allowed for better technology in private systems such as SONET, DDN, Internet, VPN's. The fiber optic Atlantic cable was an improvement in telecommunications transmission media and signal integrity.

The two federal government agencies responsible for telecommunications are the FCC and GSA. GSA contracts the government's telephone companies. FCC writes the regulations that the telephone industry must follow. The contract for telecommunications for the government was a ten year deal. The person monitoring the contract is always someone very famous in the IT community. The GSA usually bids the project and gets the best deal for the government that they can from the lowest bidder. The GSA are experts in the daily business lines of running the government. They also have an office that buys government real estate and manages Information Technology which are two other critical government infrastructure issues. They are impacted by the laws, rules, and regulation set forth by the Congress. GSA has authority to help setup contractors for the military telecommunications systems also.

Since most long haul telecommunications with data is done in encrypted binary code table 1 shows binary, octal, hex, and decimal numbers. Computers and switches talk in all of these codes using ASCII and EBCDIC coding schemes. ASCII has 128 characters as shown in Appendix E and is used on microcomputers and minicomputers mainly. Mainframes use IBM's EBCDIC code set as listed in Appendix F. The numbers represent data at the physical layer in the OSI model when transmitted. Computer convert the numbers to their system when they receive the binary data after transmission.

Modern Communications Systems

Table 1. Numerical Systems

Decimal	Binary	Octal	Hexadecimal
0	000	0	0
1	001	1	1
2	010	2	2
3	011	3	3
4	100	4	4
5	101	5	5
6	110	6	6
7	111	7	7
8	1000		8
9	1001		9
10	1100		A
11	1011		B
12	1100		C
13	1101		D
14	1110		E
15	1111		F

Some of the encryption done in communications involves shifting the ASCII code set characters by a fixed number of positions using an algorithm. An algorithm with prime numbers as the divisor or multiplier is perfect since primes are non- reducable. This would be the substitution type of ciphering. If we start with the 128 ASCII character set and shift the bits by an offset of prime numbers then we have a new value for the ASCII codes 1-128 and can transmit using this new code set. Multiple ASCII code sets can be produced with multiple prime number algorithms, one for each day of the year and we have 365 different code sets encrypted using the ASCII code set we started with. The question is how easy would this be to determine or decrypt by an enemy or interceptor of our electronic message.

Modern Communications Systems

5. Military and Law Enforcement Communications

Military communications was required for winning wars. Here are some examples of military communications that revolutionized communications. Military communication can include a variety of devices and many that are embedded into weapons systems like aircraft, tanks, ships. Some of these technologies include mounted sonar, radar, RF communications, microwave, satellite communicators, towed array phase radar, sonar buoys, electronic jammers, UHF and VHF, closed circuit TV, aircraft radio navigation, harbor navigation, commercial aircraft, wireless cable, cellular telephone and radio, and shortwave radio. Many of these signals are encrypted and ciphered so that a potential enemy can not eyes drop on the message. The computer offers a great resource to accomplish these tasks. Interestingly enough local police and state police communication systems are very similar to military communications requirements with emphasis on global reach. They communicate on the 700-800 Mhz frequencies locally. Figure 8 shows the US military alphabet learned to ensure there are no errors on telephonic communications. A good book on military contracting including communications systems is called the History of Government Contracting by Nagle from the George Washington University National Contract Law Center. It outlines the government contracting for the military going back to the continental congress and the cannonball committee. Today's military hardware is bought on contract from private defense companies.

Figure 8 . US Air Force Communications Alphabet

Alpha -   A
Bravo -   B
Charlie –   C
Delta -   D
Echo -   E
Foxtrot -   F
Golf -   G
Hotel -   H
India -   I
Juliet -   J
Kilo -   K
Lima -   L
Mike -   M
November – N
Oscar -   O
Papa -   P
Quebec –   Q
Romeo -   R
Sierra -   S
Tango -   T
Ulysses -   U
Victor -   V
Whiskey -   W
Xray -   X
Yankee -   Y
Zulu -   Z

Morse Code - Samuel Morse created Morse Code as a group of symbols of dots and dashes to communicate through tapping or electronic signals.

Samuel Morse developed Morse code for communicating by the teletype. Table 2 shows the Morse Code codes that Morse gave to the industrial revolution and which was used by many industries including the government run rail roads systems and post offices during the first industrial revolution of the late 1800's. It is interesting that American soldiers who were held prisoners in Hanoi used Morse Code to talk to each other through the walls of the prison.

Table 2. Morse Code Telegraph Characters (o = dot, - = dash)

A o -	T -
B - o o o	U o o -
C - o - o	V o o o -
D - o o	W o - -
E o	X - o o -
F o o - o	Y - o - -
G - - o	Z - - o o
H o o o o	, - - o o - -
I o o	. o - o - o -
J o- - -	1 o - - - -
K - o -	2 o o - - -
L o - o o	3 o o o - -
M - -	4 o o o o -
N -o	5 o o o o o
O - - -	6 - o o o o
P o - - o	7 - - o o o
Q - - o -	8 - - - o o
R o - o	9 - - - - o
S o o o	0 - - - - -

Ancient Egypt – Ancient Egypt communicated by papyrus which was a written word on a piece of cloth or early paper. They also wrote hieroglyphics on the walls of their tombs depicting life scenes. They used an encryption wheel that allowed them to decode messages.

DDN – The Defense Data Network is the DOD telephone network. You can call anyone in the network by finding the base locator and using the public phone number. The DDN was designed for defense workers only.

Network Centric Warfare – Network centric warfare is using computer networks to direct the acts on the battlefield. Laptops and mobile computers are used on the battlefield to accept orders using networks and coordinating procedures via integrating ground, air, and sea resources.

Electronic Battlefield – The management of battlefield data through a centralized source. The concept is a unified coordination of all assets on the battlefield at the same time controlled by computers. Lockheed Martin and other defense companies such as Northrup Grumman, and Raytheon, are leaders in this field. Dave Adamy outlines the Tactical Battlefield in his book EW 103 Tactical Battlefield.

DARPA – Defense and Research Projects Agency. The agency who invented Internet in 1969 and many other useful technologies for the DOD. Many have been adopted by the private sector and public.

Navaho Code Talkers – The Navaho Indians in the Marines Corps speak in their native tongue on the radio during the Pacific war with Japan to confuse the listeners on what was being said. The Japanese never broke the code although we broke theirs several times to be in the right place at the right time including the decisive battle of Midway.

German Enigma Machine – The German Enigma Machine is captured by the Polish and delivered to the British for intercepting messages at Bletchley Park near London. This gave the allies a great advantage in the war. Enigma works

Modern Communications Systems

based on a number of rotars that turned and gave a unique code. One is on display at the Historic Electronics Museum in Linthicum. One of America's best rotary encryptors was the M-209. It was similar to rotar operation and design of the Enigma but was never captured during World War II.

Avionics – Normal avionics systems require the aircraft to communicate with external sources to direct the aircraft and operate it safely. Aircraft communications and navigation frequencies are included in the frequency spectrum chart in this book. Subsystems include navigation, data communications, the black box data recorder, and flight controls.

Surveillance Systems – Video cameras and closed circuit TV's are now used on our ships, planes, tanks, and buildings to deter crime and serve the purpose of visually verifying what once took a human to do. Verifying a missile launching on a ship's deck no longer requires a sailor to stand there and signal the bridge. We can now remotely detect these activities from a TV console.

Shortwave Radio – The United States military uses short wave radio frequencies and anyone can buy a book on shortwave and listen on a small unit bought from Radio Shack. This can teach one the frequencies and multiple uses we have for shortwave radio. One reason it is good is that the signal propagates a long distance around the world bouncing inside the ionosphere.

Modern Communications Systems

Iridium Satellite Telehones – Past presidents have used these secure telephones to communicate around the globe. Other diginatries may also use these phones. During combat operations commanders use the Iridium. The signal uplinks to the satellite transponder for repeating to the next location either ground or satellite in the Iridium network.

Radar – These systems include the British Original, TACAN, LORAN, radar jammers and detection, radar guided missiles, stealth anti-radar designs for aircraft (F-117, B-2 Spirit) and ships (Sea Shadow System). The radar range oven was the first microwave oven developed after the microwaves from radar systems were found to pop popcorn next to microwaves from the heat generated as a spinoff technology.

Sonar – These systems include Dipping Sonar Buoy, Fixed Ship Sonar (Bubble Underneath the bow), Submarine Sonar – Pings, sonar Maps of the ocean floor for research. The basic theory is that sound travels through the water and can detect the presence of objects in the water. Sonar is used in Anti-submarine Warfare (ASW), Submarines, and many other ships on the sea, Deep Sea Fishing, and Oceanographic Research.

Navigation Systems – GPS has changed the way we navigate around the world. The US Coast Guard operate a set of GPS satellite systems around the globe and ships, planes, boats, trucks and other vehicles can determine exactly where they are on the surface of the earth with respect to longitude and latitude. GPS is universal in appeal. It can even be used to track hiking expeditions in the local woods. Dead reckoning is using an object on shoreline or land to guide one's craft. A sextant is a device that allows one to navigate by the stars. Mathematics such as trigonometry helps in navigation by maps and hand calculations to determine azimuth and back azimuth's for direction of travel compared to a navigation map.

Missile Control Air Defense – NORAD is well known for automation. Computers controls the targeting of MIRVs (Multiple reentry vehicles) in the warheads for many kills on one launch. Our missile defense system is not perfect with several close calls during the cold war. That we have not had to use it is a testament to the policy of nuclear détente between nations which assures mutual destruction. The troubling thing is that we now have 17 countries with nuclear weapons systems or weapons of mass destruction. This also means they have the capacity to produce nuclear power. The Missile Defense Agency (MDA)[3] is completing America's Anti Ballistic Missile System.

Military Air Traffic Control Systems – Air Traffic Control Systems follow military flight regulations which are more strict than the FARS – Federal Aviation Regulations Operations by the Navy and Marines Corps are both sea based and land based. The Air Force operates off land bases only. The ATC job is tiring and requires patience and endurance as it is highly stressful. Most air base

---

[3] See internet website at http://www.mda.mil

towers have the controllers tracking military aircraft using computer based consoles with sweeping radar.

Fire Control (Aegis Integrated Ship Defense System) – USS Ticonderoga Class Cruisers, Arleigh Burke Class Destroyers, and now Frigates, tested at Pax River in the early 1990's. Automated radar detection and fire control of weapons systems was featured in the Phallanx gun system. Rail guns were experimented with. Aegis is Greek in origin and means "a sheild borne by Zeus, Athena, or Apollo" in mythology or "a protection" from the Webster Collegiate dictionary. The system uses advanced sensors, satellites, computers and miles of cables to form the protective shield around the ship from air, sea, or land attack weapons. The systems is the crux of the network centric warfare defense systems for the modern naval fleet, but is only one component of the overall system at the geo-commanders disposal on the new theater level battlefield. The tiome a sensor can detect and direct automated computer missile and gunfire is immediate compared to human aimed weapons. The problem is the human intervention that may make a mistake in the system identification of friend or foe. This was what happened with the Vincennes shot down of the Arab airliner in the Mediterrainian Sea. The defense system was too good and accidentally shot down civilians which were determined a threat to the ship. Aegis cruisers are a sub-component in the MDA.

Modern Communications Systems

Websites – Internet websites are as common as any systems today. The military uses them for many business applications such as recruitment, retention, personnel, medical histories (also on smart cards), education (distance format), and contract management. Terrorists use websites to organize. The FBI website is very informative for Homeland Defense.

MILNET – A military network of communications devices uniting the Pentagon and military command structures of the Department of Defense. DefenseLInk is also a systems they run.

BITNET – A university system that allows users (teachers, professors, and researchers) to communicate messages via computer systems.

The following communications systems are law enforcement and EMS communications systems that I have heard about while on duty in Washington DC.

#77 – This system is a cellular phone emergency systems that was recommended by a non technical police officer with the Maryland State Police to help people carrying cell phone in their cars when they sight an emergency or problem on the highway. It is a great way to get customer inputs from the highway system to help police ourselves.

Modern Communications Systems

911 – This systems is a local system with to police forces in the counties. The 911 call is accepted and routed to the proper person by the dispatcher and staff at the local law enforcement headquarters. Time saved in this system is life saved.

NCIC – The National Crime Information Computer has been in operation at the FBI headquarters for a long time. They have a display at the headquarters building for visitors to view when they are on the FBI tour of the building. The NCIC allows a local law enforcement agency to send for information to FBI headquarters and have it sent back to the field level in the officer's vehicle in 309 seconds or less. That's an amazing response time. But it does not cover all information from all jurisdictions.

CAPWIN – The CAPWIN project was created after the attacks on New York City and the Pentagon and the purpose is to link local police, fire, and EMS aganicies together for resources needed between local jurisdictions rather than having them always work through the FBI and NCIC systems and other systems. Some data needed by law enforcement and during emergiencies is stored in local databases and needs to be transmitted to sister local agencies such as Maryland MVA to Virginia police officers checking a Maryland titles or Maryland arrest records on the Capital beltway.

Modern Communications Systems

CB Radio – Channel 9 is the police emergency channel of the Citizen's Band Radio. Truckers mainly use Channel 19. An emergency can be reported to the authorities on Channel 9 using any typical CB radio. There is a protocol for talking in the CB radio and you should have your "handle" ready when talking to others while driving.

Fax systems – Faxes are a way of life in America in all walks of the business world. Fax machines are able to send visual signals and similes. The fax has improved the way business images are transmitted. Contract proposals and important legal documents are transmitted this way. Fax wars are when one company ties up another companies fax machines so that that company can not send a proposal on a project that is important to the business botoom line of the one company. A kind of high tech fax jamming as it were to stymie the competition. If the fax is busy receieving junk mail it can not send out any valid proposals that may take business away from the other company.

Voice Messaging – Voice messaging systems are very abundant these days. The types of systems are great. Many hold more than 30 messages electronically by turning the voice into a signal stored on the machine. These machines are usually attached to the telephone system and ring over when the telephone is not answered. Another common name for them is answering machines. In business, the answering machine and voice mail have become a necessity to receive all incoming calls when not at one's desk.

Modern Communications Systems

VOIP – Voice Over Internet Protocol is piggybacking the internet signal with a voice signal. Magicjack (NYSE ticker CALL) does this on the home PC. It is a small device for $49 that makes home telephone service free. It is and ingenius money saving device. The bad news is that you have to run your PC all the time to keep the Magicjack active. There is software loaded on the PC that installs wjhen the computer starts up. You keep you regular phone number. It saves a lot of money. The author currently ruins this on Version FIOS through a wireless router.

Electronic Warfare – Association of Old Crows teaches EW. They are helpful in learning all aspects of EW for military and non-military people. The military has this as an MOS for staff. Dave Adamy is the best EW teacher.

Information Warfare – the use of information for pruposes of destroying an enemy networks and compurer resources. We live in the computer age and this has become an essential part of both intelligence operations and military operations.

Cyberwarfare – made famous by an author named Richard Clark. He was a previous advisor to president Bush in Homeland Security. He cited cybersecurity as one area of focus by our enemies. It escalates to cyberwar when real fighting breaks out after a cyber attacks. A nuclear attack after a cyber attack would be an example of this. North Korea and Iran has threatened both of these in the past.

6. Packet Switching Networks

Packet switching networks include Internet and any private networks that use switches to "store-and-forward" messages to other switches and computers in a network. A switch is a computer controlled routing device for messages that contains an address and destination tables. A packet is the basic unit for messages in the packet switching network. It has headers and trailers in it as well as data that is the message itself. The two types of packet switching are the Virtual Packet and the Datagram Packet. The Virtual Circuit packet travels along the same path of switches to the destination. The datagram packets travel on any path of switches to the destination. The datagram is a type of packet switching message that can be transmitted on a packet switching network.

Figure 9. Virtual Circuit Packet Route

```
Source -----→ SW 1        SW2           SW7
               |
               v
              SW3---------------→SW4--------→SW5------→ Destination
```

Figure 10. Datagram Packet Routes

```
Source-------→SW1-------→SW2------------→SW7----------------------
               |                                              |
               v                                              v
              SW3-------------→SW4------------→SW5-----→Destination
```

The TCP/IP protocol is based on packet switching. Satellites can send packet switched messages to the next hop. The destination address and source address of the message are both contained in the packet. Switches and routers have tables that contain the next address of the next closest switch to the destination in the network. Network analysis tells the router or switch what the best route to take (least congested) at any one time. The biggest secret about packet switching is that the largest messages can be broken down into small messages and sent in any order on the switched network and reassembled at the destination address. The network does not care how the messages are sent only that they are marked with a sequence number so they can be put in order before being read by the final receiver. Forouzan has many good illustrations of how packet switching networks work in his outstanding technical books from Prentice Hall.

Since Internet was developed by DARPA it was thought it was to be used as a military system only. This meant that it had to be very secure and survivable in the design of the system. Multiple switches were separated by great distances to ensure that a nuclear attack would not totally knock down all switches in the network. The very fact that a message is broken up when it is transmitted and sent by various paths to the end destination was a survivable design specification of internet. The man who kept all the Internet domain names passed away a few years ago and passed his job on to another.

A new packet switching network known as "Internet in the Sky" is under development by Microsoft and TeleDesic company. The project is delayed but started in 1996 and is scheduled for 2005 completion. The idea is to use all satellites to send messages around the world. The gain is speed of the messages sent through the network. It is not clear if this is the same as Internet2 which is also a project under development. If Microsoft is allowed to develop and make major investments in the network, then you can be sure it will be compatible with their products. This further enhances their monopoly on the Windows operating systems market which runs on many types of manufacturer's hardware.

A PBX is a Public Branch Exchange and is a type of switch for a large building or city block of telephone addresses. A PBX can be linked to a LAN system where telephone (voice data) can be integrated with computer data. PBX systems are generally installed when the building is installed and often are found in a wiring closet on the older buildings. Newer buildings are smart buildings and have their own wiring in the building that gives it the "smart" characteristics.

Modern Communications Systems

7. The ISO OSI Model

First the ISO means International Standards Organization. They also have standards on Quality such as the ISO 9000 standards and they are recognized around the world not just in America for many standards. It is an honor when an organization is ISO compliant on any standards. Private companies usually benefit from the standards more than government organizations but when contracting for products and services it can make a difference to the government too.

The Seven Layer OSI Model

The OSI model has 7 layers that every telecommunications student memorizes before learning the specific hardware and software he needs to know to build a network or manage a network. The layers are application, presentation, session, transport, network, data link, and physical layers. They are formed into sender and receiver layers. Figure 11 shows the basic idea of the open systems interconnection model.

Figure 11. The Seven Layers OSI Model

Seven layers are defined:

7) **Application** : Provides different services to the applications

6) **Presentation** : Converts the information

5) **Session** : Handles problems which are not communication issues

4) **Transport** : Provides end to end communication control

3) **Network** : Routes the information in the network

2) **Data Link** : Provides error control between adjacent nodes

1) **Physical** : Connects the entity to the transmission media

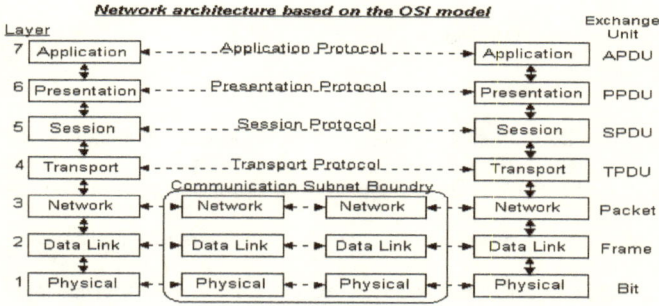

Source: **http://www2.rad.com/networks/1994/osi/layers.htm** accessed on
16 December 2003 at 1630

Layer-Specific Communication

Each layer may add a Header and a Trailer to its Data (which consists of the next higher layer's Header, Trailer and Data as it moves through the layers). The Headers contain information that specifically addresses layer-to-layer communication. For example, the Transport Header (TH) contains information that only the Transport layer sees. All other layers below the Transport layer pass the Transport Header as part of their Data. Figure 12 shows the physical layer PDUs that form the basis of the frames sent through the OSI model.

Figure 12. Physical Layer PDU's

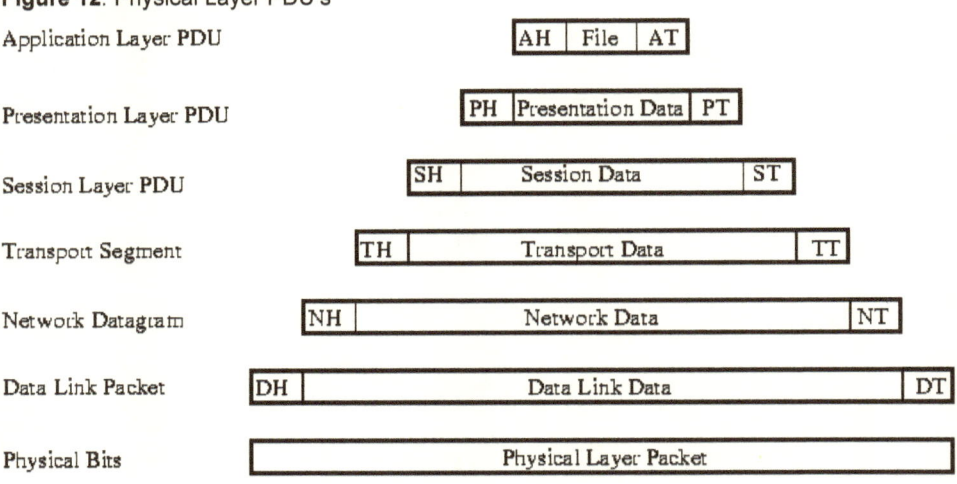

PDU - Protocol Data Unit (a fancy name for Layer Frame)

Source: https://secure.linuxports.com/howto/intro_to_networking/c4412.htm accessed on 16 December at 1700 hours.

Modern Communications Systems

Figure 13 shows the vendor products relationships of protocols to the OSI model. This is important because you have to be able to integrate protocols in network systems and software layers using various manufacturers. The OSI model gives you a pictoral tool to memorize how the layers are interconnected. But it is only a start.

Figure 13. Protocol Stacks in Relationship to the OSI Model

OSI Layer	Apple Computer	Banyan Systems	DEC DECnet	IBM SNA	Microsoft Networking	Novell NetWare	TCP/IP Internet	Xerox XNS	OSI Protocols
Application Layer 7	Application Programs and Protocols for file transfer, electronic mail, etc.								
Presentation Layer 6	AppleTalk Filing Protocol (AFP)		Network Management Network Application / Remote Procedural Calls (Net RPC)	Transaction Services Presentation Services	Server Message Block (SMB)	NetWare Core Protocols (NCP	(Telnet, FTP, SMTP, etc.)	Control and Process Interaction	ISO 8823
Session Layer 5	AppleTalk Session Protocol (ASP)		Session	Data Flow Control	Network Basic Input/Output System (NetBIOS)	Network Basic Input/Output System (NetBIOS)			ISO 8327
Transport Layer 4	AppleTalk Transaction Protocol (ATP)	VINES InterProcess Communications (VIPC)	End Communications	Transmission Control	Network Basic Extended User Interface (NetBEUI)	Sequenced Packet Exchange (SPX)	Transmission Control Protocol (TCP), Unacknowledged Datagram Protocol (UDP)	Sequenced Packet Protocol (SPP)	ISO 8073 TP0-4
Network Layer 3	Datagram Delivery Protocol (DDP)	VINES Internet Protocol (VIP)	Routing	Path Control		Internet Packet Exchange (IPX)	Internet Protocol (IP)	Internet Datagram Protocol (IDP)	ISO 8473 (CLNP)
Data Link Layer 2	Network Interface Cards: Ethernet, Token-Ring, ARCNET, StarLAN, LocalTalk, FDDI, ATM, etc. NIC Drivers: Open Datalink Interface (ODI), Network Independent Interface Specification (NDIS)								
Physical Layer 1	Transmission Media: Twisted Pair, Coax, Fiber Optic, Wireless Media, etc.								

Source: http://www.lex-con.com/osimodel.htm accessed on 16 December at 1600.

Modern Communications Systems

Physical Transmission

Usually, data bit streams are sent in groups of 4 bits with headers and trailers to the basic message format. Many authors describe the ways this basic transmission of physical data bits occurs in frames, messages, and the types of systems that do the functions by various proprietary systems such as Forouzan[4] and Panko[5]. The idea is to understand the model well enough to use it during ones work in today's computing environments and networks on various systems engineering projects. Some authors like Stallings[6] invalidate the importance of the OSI model in favor of TCP/IP, but knowing the diagram in figure 13 can help you build a network in those proprietary protocols like DECNET and Novell as network administrator.

A metaphor of the generals

Suppose two generals want to talk to each other. The sending general wishes to ask another general in another military unit to dinner. The sending general asks his colonel to translate the message and give it to the major and captain and so on down to a private and then the private gives the message to the other unit's private and he gives it up line to his corporal, sergeant, lieutenant, captain, major, colonel, and finally the receiving general. The receiving unit

[4] Forouzan, Data Communications and Networking, McGraw Hill, 2004.

[5] Panko, Ray, Business Data Networks and Telecommunications, Prentice Hall, 2003.
[6] Stallings, William, Business Data Communications, Prentice-Hall, 2005.

decodes it at each level and send it up the organization. The receiving general reads the message and says thanks "I Accept the Dinner Party with General A". This metaphor works for any two organizations where the message needs to be encoded, transmitted securely, and decoded on the other side.

Figure 14. The Metaphor of the Generals

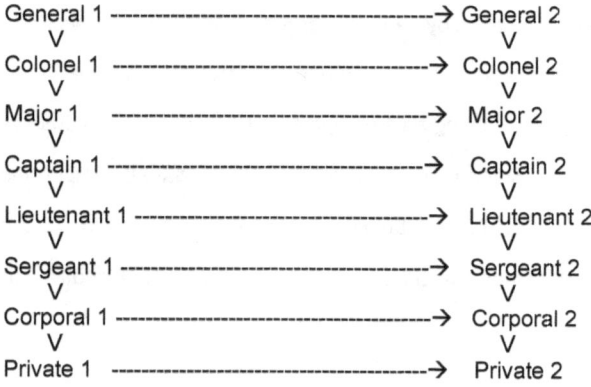

Modern Communications Systems

8. Radio Propagation and Full Spectrum Frequency Management

Signals Background Knowledge

Radio signals move at the speed of light. Marconi invented the modern radio. The US Army Radio Propagation Manual describes how signals are skywave, space wave or ground wave. We have images of these below. The Earth has several layers of atmosphere where signals are transmitted. At night two of these layers (D & E) combine to make one layer. This makes reception of SW radio easier. Ham Operators use HF signals in similar fashion to AM/FM/SW. The AM/FM/SW signals waveforms are shown below. There is also a comparison of analog and digital waveforms. Analog wave forms are a sine and cosine wave pattern. Fourier Series describe the many patterns a signal may take mathematically. The digital signals are square signals with bits of 1s and 0s shaping upper and lower squares. Conversion from analog to digital waveforms is required in a CODEC. Digital signals are used in computer communications on internet. The digital signals are also used in Satellite transmissions. These satellite are stationed above the horizon at various KM heights above earth to cover the earth in geosynchronous orbit (Ka Band). A ground station uplink is needed to connect the signals to the satellite. Recently DOD research has led to Software Defined Radio which uses the PC computer to scan all frequencies in the radio spectrum with higher quality signals and more accuracy.

Figure 15. Frequencies Bands

Band	Frequency range	Wavelength range
Extremely low frequency (ELF)	< 3 kHz	>100 km
Very low frequency (VLF)	3 - 30 Hz	10 - 100 krn
Low frequency(LF)	30 - 300 kHz	1 - 10 km
Medium frequency (MF)	300 kHz - 3 MHz	100m - 1km
High frequency (HF)	3 - 30 MHz	10 - 100m
Very high frequency (VHF)	30 - 300 MHz	1 - 10m
Ultra high frequency (UHF)	300 MHz - 3 GHz	10cm - 1m
Super high frequency (SHF)	3 - 30 GHz	1 - 10cm
Extremely high frequency (EHF)	30 - 300 GHz	1mm - 1cm

Figure 16. Shortwave Frequencies

Metre Band	Frequency Range	Remarks
120 m	2300–2495 kHz	tropic band
90 m	3200 – 3400 kHz	tropic band
75 m	3900 – 4000 kHz	shared with the North American amateur radio 80m band
60 m	4750 – 5060 kHz	tropic band
49 m	5900 – 6200 kHz	
41 m	7200 – 7600 kHz	shared with the amateur radio 40m band
31 m	9400 – 9900 kHz	currently the most heavily used band
25 m	11,600 - 12,200 kHz	
22 m	13,570 - 13,870 kHz	
19 m	15,100 - 15,800 kHz	
16 m	17,480 - 17,900 kHz	
15 m	18,900 - 19,020 kHz	almost unused, could become a DRM band
13 m	21,450 - 21,850 kHz	
11 m	25,600 - 26,100 kHz	may be used for local DRM broadcasting

Pertinent Images, Charts, Diagrams

The following images and charts were discovered after research on signals analysis, RF propagation, Fourier Series mathematics, and Satellite digital communications. Each diagram takes the reader from basic analog to advanced digital communications. Though our study was AM/FM/SW/SDR we also use digital communications every time we log into a computer and send a message via satellite or other digital network like Ethernet, Token Ring, SNA, TCP/IP or OSI model. The path of the digital communications is not predetermined if it is computer packet switching. That means a signal can be coded into digital and sent along any type of path from sender to receiver (even satellite or fiber optic) without knowledge of the end user.

Figure 17. D,E F1, F2 Layers of the Atmosphere.[7]

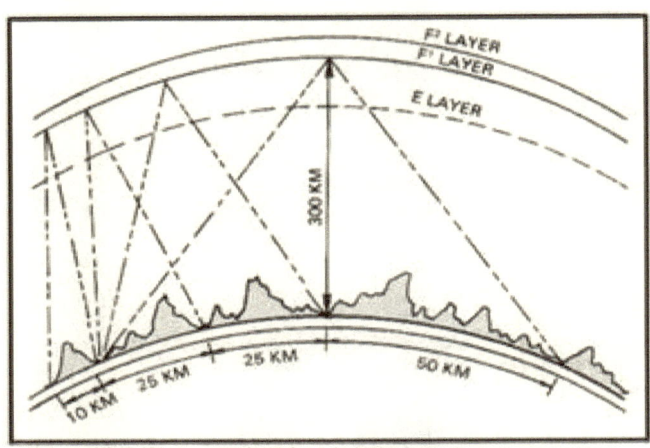

Figure M-1. Near-vertical incidence sky-wave propagation concept.

---

[7] US Army Radio Propagation Manuals, 1986 & 2009, US Army.

Figure 18: Ionosphere skywave signals around earth.

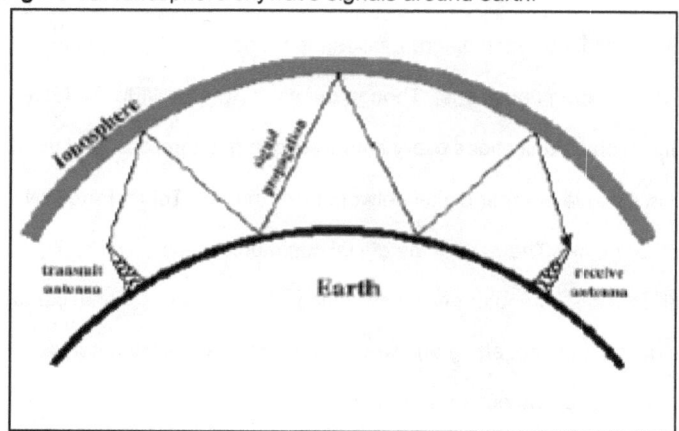

Figure 19. Line of Sight between antennae versus Skywave

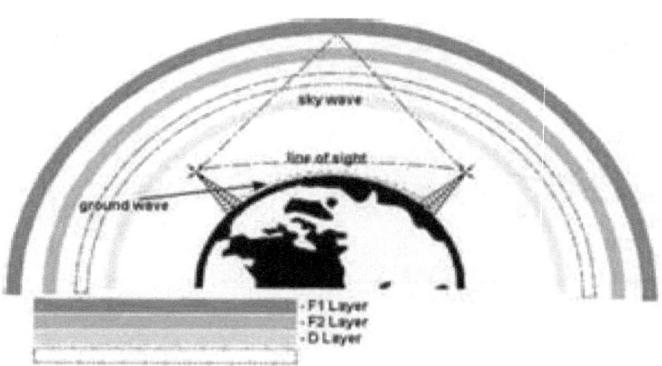

Figure 20. Skywave across ocean. Space wave. Surface wave.

Figure 21. Skip Zone and Skip Distance of Skywave.

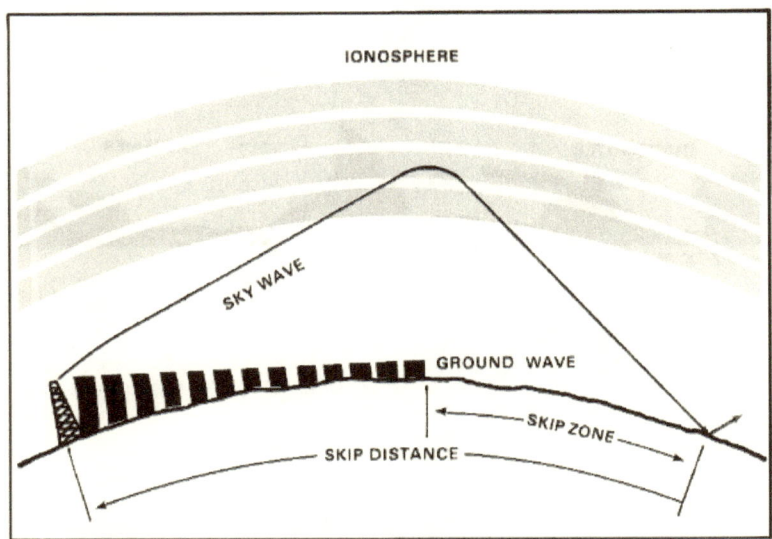

Figure 2-15. Low angle sky-wave transmission path.

Modern Communications Systems

Figure 22. Sinusoidal Waveform Cycle

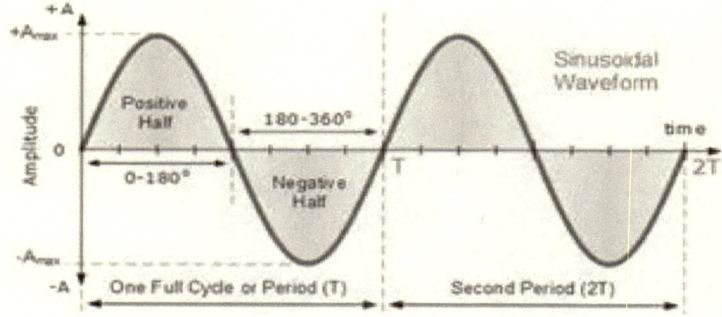

Figure 23. One cycle of analog wave or one hertz.

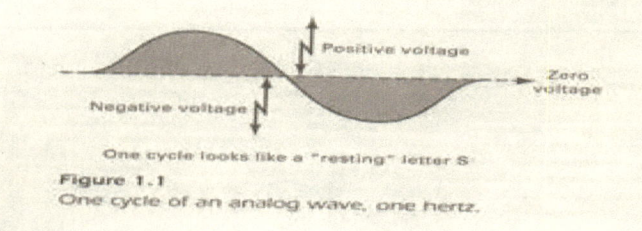

Figure 24. AM and FM comparison waveforms

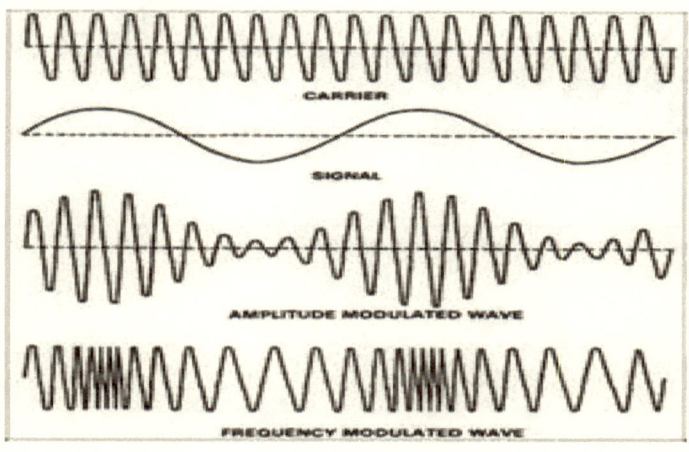

Figure 25. Analog and Digital Signal

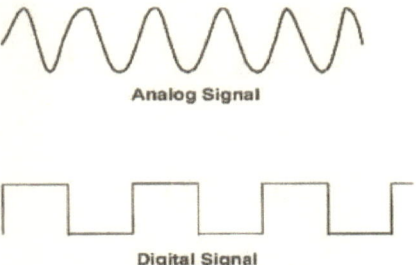

Figure 26. CODEC Analog - Digital Converter

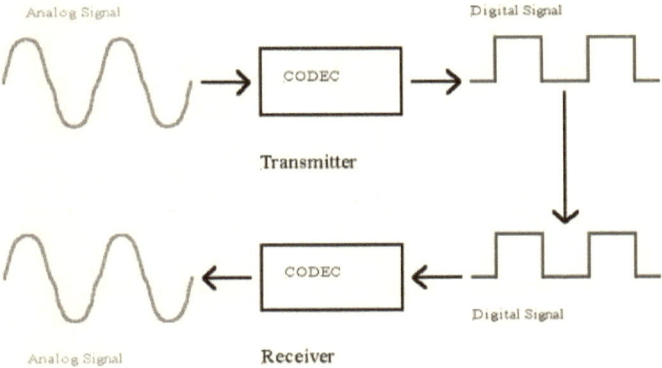

Figure 27. Other Fourier Series waveforms and math equations[8]

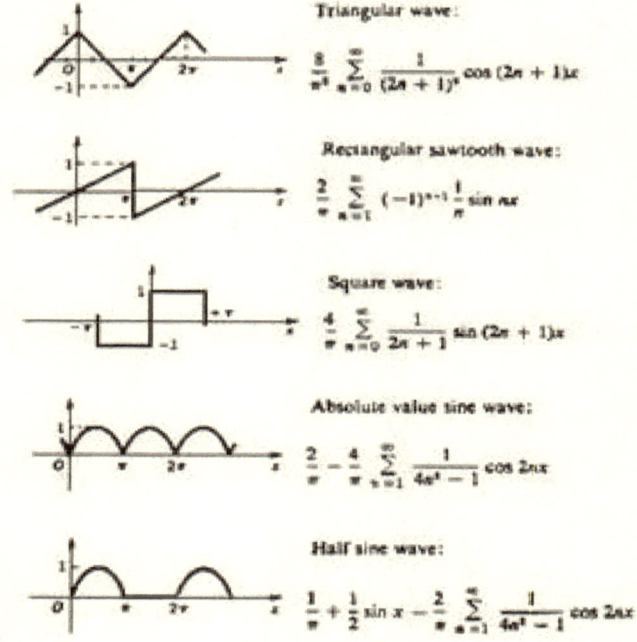

Triangular wave:

$$\frac{8}{\pi^2} \sum_{n=0}^{\infty} \frac{1}{(2n+1)^2} \cos(2n+1)x$$

Rectangular sawtooth wave:

$$\frac{2}{\pi} \sum_{n=1}^{\infty} (-1)^{n-1} \frac{1}{n} \sin nx$$

Square wave:

$$\frac{4}{\pi} \sum_{n=0}^{\infty} \frac{1}{2n+1} \sin(2n+1)x$$

Absolute value sine wave:

$$\frac{2}{\pi} - \frac{4}{\pi} \sum_{n=1}^{\infty} \frac{1}{4n^2-1} \cos 2nx$$

Half sine wave:

$$\frac{1}{\pi} + \frac{1}{2} \sin x - \frac{2}{\pi} \sum_{n=1}^{\infty} \frac{1}{4n^2-1} \cos 2nx$$

Figure 28. Communications Satellite Space Wave propagation

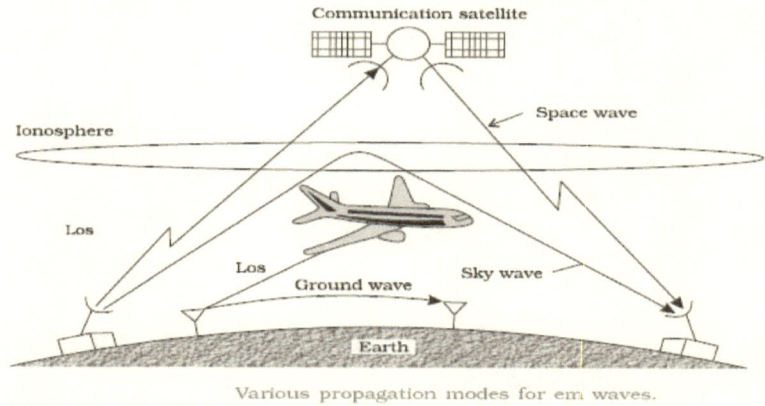

Various propagation modes for em waves.

[8] Tolstoy and Silverman, Fourier Series, Dover Publications, 1976.

Modern Communications Systems

Figure 29. Satellite Dish - Focus of Parabola From Trigonometry & Calculus

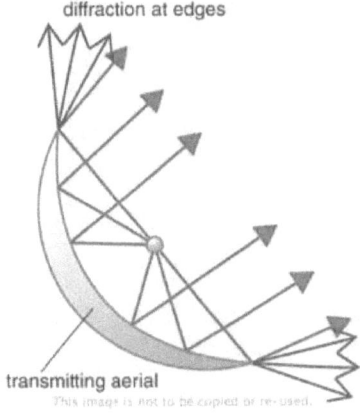

diffraction at edges

transmitting aerial

This image is not to be copied or re-used.

Figure 30. Internet Satellite Communications - Digital

Figure 31. Satellite Receiver Circuit[9]

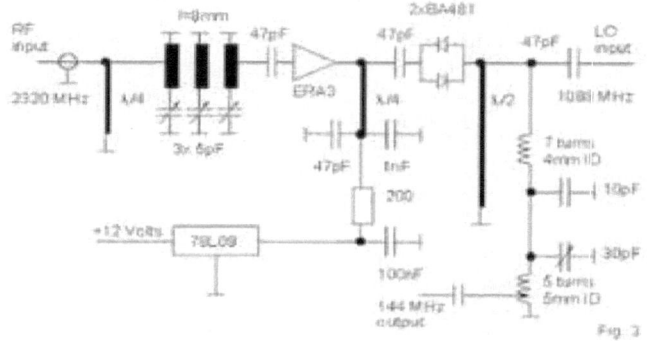

Figure 32. Radar Waveforms

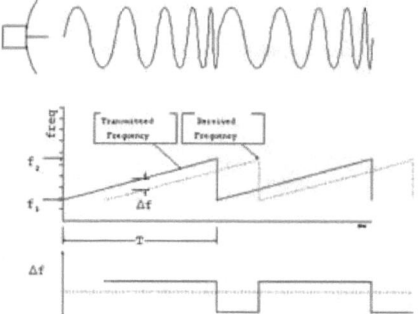

Figure 33. Radar Pulse Signal Width in One Direction Towards Target (versus Broadcast like AM, FM, SW, SDR)

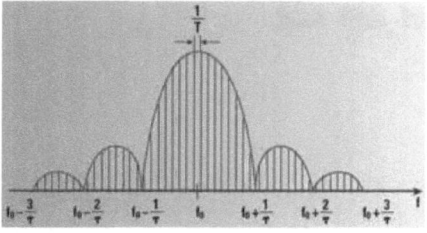

[9] Graf, Encyclopedia of Circuits, Vol 7, 2009.

Figure 34. NASA Glenn Research Center Deep Space Ka Band RF Propagation

Stations around the Earth

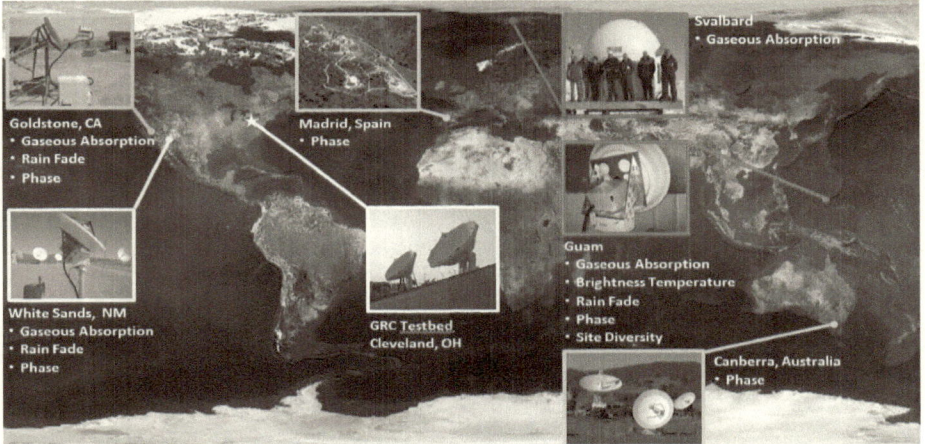

EE Communications Scientific Theories at work

Electro magnetism (Faraday and Maxwell) - book on the life of both of them was browsed. They lived in a time when chemistry was becoming physics and Newton was king. Einstein changed everything with Theory of Relativity and curvature of space time continuum.

RF Radio Propagation - gleaned from US Army Radio Propagation Manuals. Earth Atmosphere RF Characteristics - Gleaned from US Army Radio

Modern Communications Systems

Propagation Manuals and diagrams on internet.

Information Theory (*Communications Theory*) - mathematical explanation of how we communicate using RF and computer. Once known as general Communications Theory it is now a field of the IEEE.

Analog to Digital Signal Conversions - diagrams available on this. See figure 26. This is critical because one must understand both analog and digital waveforms today in communications and computers.

Digital Transmission - square wave forms also a part of Fourier Series mathematics. Bits of 1s and 0s sent across channels in square waves rather than sine or cosine analog waves.

Analog and Digital Circuit Boards - schematics available for AM/FM/SW/satellite from Encyclopeida of Circuits Vol 7 by Graf. .

SDR – Software Defined Radio – a way that DOD researched in the 1990s to adapt the PC computer to scan all frequencies more efficiently than any other regular radio circuits. Software drives the radio frequency searching. Tested by author in EE class project.

Modern Communications Systems

Digital Software Protocols (TCP/IP, OSI Model) - from work on previous networks and data communications projects by author. TCP/IP or internet protocol and Open Systems Interface 7- layer model.

Radar signals waveform - a pulse for pulse doppler with sidebands towards a target and reflected back to receiver.

NASA - is studying RF propagation at the highest frequencies in deep space communications for the last 20 years. The ground stations in figure 34 show where they are located on the planet. Ka Band deep space communications is being utilized at the upper end of the Frequency Spectrum.

Full Spectrum Frequency Management

It was once said that communications management is frequency management of all communications and is device independent. FCC would be very involved with the creation of these charts and management of the airwaves. Figure 35 is an abbreviated chart of the frequency band ranges from Panko[10]. This figure clearly shows the types of uses of the various frequencies of the spectrum from Extremely Low Frequencies to Extremely High Frequencies. Some of the frequencies are used on our daily communications devices such as TV's and radios. Others are used in higher cost technologies such as satellites, microwaves, and wireless LANs. The FCC regulations covering communications is Title 47 of the Code of Federal Regulations. FCC manages the industry using

---

[10] Panko, Business Data Communications and Networks, pp. 424.

these regulations as guides and deregulation to invoke more competition such as

the 1984 Telecommunications Deregulation Act.

Figure 35. Full Spectrum Frequency Bands Ranges

Band	Full Name	Uses	Lowest Freq	Bandwidth	Units	Wavelength of lowest Frequency	Units
ELF	Extremely Low Freq		30	270	Hz	10,000	Km
VF	Voice Frequency	Telephone	300	2,700	Hz	1,000	Km
VLF	Very Low Frequency		3	27	kHz	100	Km
LF	Low Frequency		30	270	kHz	10	Km
MF	Medium Frequency	AM Radio	300	2,700	kHz	1,000	m
HF	High Frequency		3	27	MHz	100	m
VHF	Very High Frequency	VHF TV, FM Radio	30	270	MHz	10	m
UHF	Ultra High Frequency	UHF TV, Cellular Phones	300	2,700	MHz	100	Cm
SHF	Super High Freq	Satellites Microwaves Wireless LANS	3	27	GHz	10	Cm
EHF	Extremely High Freq	Future Q/V Band Satellites	30	270	GHz	10	Mm

Modern Communications Systems

The chart in figure 36 is the detailed frequency spectrum chart that can be used for managing the spectrum of services. This list was derived from CED 2003-2004 Frequency Allocations from the internet.

Figure 36. Detailed Frequency Spectrum by Function

Name / Function	Frequency Range (MHz)
Shortwave Radio	7-26
10 Meter Ham	28-29
Land Mobile	42-50
Ham	50-54
VHF TV	54-88
FM	88-108
Aircraft RadioNavigation	108-117.96
Aircraft	117-121.5
Commercial Aircraft (Air Traffic Control)	122.5-136
Meterologic Satellite	137-138
Space to Earth Research	139-142
Civil Air Patrol	143
Mobile Satellite Service	144-160
Land Mobile	150-156
Maritime Mobile	156-157
Land Mobile	158-161
Government Fixed Mobile	162-173
Fixed Land Mobile	173.2-173.4
VHF TV	174-216
Fixed Land Mobile	216-220
Private	220-222
Ham	223-225
Government fixed Mobile including Aero- Communications	226-300
Harbor Navigation and Coast Guard	301-328.6
Aeronautical Fixed Naviagtion	328.7-335.4
Government Fixed and Mobile	335.5-399
Mobile Satellite Services	399.9-400.05
Medical Implants	404-405
Government Fixed and Mobile	410-420
Ham	420-450
UHF TV / DTV	470-698
Fixed and Mobile	699-750

Modern Communications Systems

UHF Channels 60-69	751-806
Special Mobile Radio	807-821
	822-824
Cellular System	825-849
Air-ground	850-851
Private Base	851-868
Public	868-869
Cellular Systems Public Base	870-890
Air-ground	894-896
Narrowband PCS	900-902
Radiolocation	903-928
Narrowband PCS	930-931
Paging Systems	932-935
Land Mobile	936-946
Private Fixed	947-960
Aircraft Radionavigation	961-1216
Radionavigation and Ham	1240-1300
Aeronautical radionavigation	1301-1360
Fixed and Mobile	1380-1382
Medical telemetry	1427-1429
Telemetry	1430-1432
Aeromedical telemetry	1433-1469
Mobile Satellite Service	1610-1626
Rand MSS	1627-1640
Fixed and Mobile	1670-1676
PCB	1860-1910
Low Power Unlicensed PCB	1910-1920
PCB	1920-1990
CARS Mobile	1190-2110
AWS	2106-2180
Mobile Satellite Services	2180-2290
Ham	2300-2310
Digital Audio Radio	2320-2346
Wireless Communications Service	2347-2380
Fixed and Mobile	2380-2390
Unlicensed PCS	2390-2400
Fixed and Mobile	2400-2410
MSS	2493-2600
Wireless cable	2600-2690
Radio Astronomy	2690-2700
Radionavigation	2700-2800
Radionavigation	2800-3000
Gov Radio Location	3000-3400
FSS	3400-4200

Modern Communications Systems

Fixed and Mobile	4200-5000
NGSO M98 header uplink	5091-5250
802.11	5470-5728
Ham	5728-6000
Intelligent Transportation Service	6000-6026
Common Carrier	6000-6400
CARS mobile	6472-6626
NGSO MSS	6700-7025
CARS Mobile	7000-7125
Satellite	8000-8000
Digital Electronic Message Service	10660-10900
NGSO FSS	10900-12300
Private Conventional Fixed Service	12700-13260
CARS	17700-18300
NGSO & FSS	18800-19300
Fixed and header for MSS	19300-19700
Domestic Public Fixed, Private Operational Fixed	21200-22600
Digital Electronic Messaging Service	24000-24460

Modern Communications Systems

9. Protocols and Communications Ports

The discussion of protocols can be confusing to students. The best way to explain them is that when one general wants to talk to another he sends the message down to a private who delivers the message and gives it to the other general. Generals relate to generals, colonels relate to colonels, majors relate to majors, captains to captains and so forth. Here are some questions and answers about protocols.

1. Why do we need protocols?

They standardize the communications process and make it easier to convert electronic messages at each level in the OSI model. This is why we need to memorize the OSI model and understand how it functions.

2. What is a communications protocol?

A way of communicating electronically at a certain levels in the OSI model.

3. What is a communications port?

A doorway on the computer to and from the network physical layer where messages are sent through a network via a protocol. It is like the entrance ramp to the interstate system.

The port numbers are divided into three ranges: the Well Known Ports, the Registered Ports, and the Dynamic and/or Private Ports.

The Well Known Ports are those from 0 through 1023.

The Registered Ports are those from 1024 through 49151

The Dynamic and/or Private Ports are those from 49152 through 65535

Appendix B is a list of frequently seen TCP and UDP ports and what they mean. The goal of this port table is to point to further resources for more information.

Figure 37. VERY WELL KNOWN PORT NUMBERS

Port Number	Function
005	RJE
011	Sysstat
015	Netstat
021	FTP - data

023	FTP
025	Telnet
027	SMTP
053	DNS
066	SQLNET
069	TFTP
070	Gopher
079	Finger
080	HTTP (World Wide Web)
088	Kerberos
102	X.400
109	POP2
110	POP3
161	SNMP

The Well Known Ports are assigned by the IANA and on most systems can only be used by system (or root) processes or by programs executed by privileged users.

Ports are used in the TCP [RFC793] to name the ends of logical connections which carry long term conversations. For the purpose of providing services to unknown callers, a service contact port is defined. This list specifies the port used by the server process as its contact port. The contact port is sometimes called the "well-known port".

To the extent possible, these same port assignments are used with the UDP [RFC768].

The range for assigned ports managed by the IANA is 0-1023.

Appendix D has some default ports used by Trojan Horses. Trojan Horses are software that enters the system and then activate based on some event after being welcomed inside the computer system firewall and outer protection defensive software layers.

Figure 38 shows some of the types of protocols you may work with in your computer applications. You may not know what protocol you are using but it is always there transparently delivering your messages to the other end of the connection. The protocol may depend on the network topology and the physical connection type.

Modern Communications Systems

Figure 38. Type of Protocols

XModem/CRC

YModem

ZModem

KERMIT

TCP/IP

SNMP

X.25

DECNET

NET8

Token Ring (IEEE 802)

Ethernet

A protocol connects the computers together and links them to be able to communicate using the binary data on the physical layer of the OSI model. The program that connects them (binds them) to the sockets in the operating system on the physical ports. Testing a new protocol requires more than one computer connected point to point. It is a flexible design in a telecommunications package that allows the end user to choose his protocol. The reaons you want to be able to pick your protocol is that some protocols were designed for batch processing of transactions (YMODEM) and some were designed with error checking in them (XMODEM/CRC). The CRC stands for Cyclical Redundancy

Modern Communications Systems

Checking at the end of the message. Some protocols were designed to be mainframe protocols (KERMIT). Some protocols were designed for specific hardware (DECNET, X.25, SNA). Some are designed for databases (SQLNET). Most protocols are layered on top of the physical connection and used in the applications software loaded on both ends of the connection or the server and client in the LAN configuration. Refer to the OSI model for general details. Hopefully, you will never have to write a protocol but only install and connect using one in a start up script or utility.

Modern Communications Systems

10. Point-to-Point Systems

The home system enthusiast has many choices in today's data communications market. He can choose to connect to a dial up ISP server, DSL line, or broadband cable modem. He can then setup a network of computers in his home through Ethernet or wireless connectors. The operating systems he can choose from include Apple OS, Windows XP, Unix, and Linux. The point-to-point setup requires knowledge on the part of the system owner to install and manage the setup. The costs are less than $50 a month as of 2004 for cable modem and $19 for unlimited internet service. This cost brings the technical expertise of the company and the monthly service. The companies are regulated by FCC laws. Various locations and counties have various cable services and ISP services available to them. Usually only one or two cable companies service a county. Many ISPs service many areas and a cable company would also be considered an ISP. Some common ISP's are listed in figure 39.

Figure 39. Internet Service Providers

AT&T World

Comcast

Earthlink

Erols

Modern Communications Systems

Verizon

America Online

Netzero

Juno

Clarknet

Cox

Modern Communications Systems

You may pay various prices for various services in figure 39, but you should try whatever service you feel will deliver the service you need. If you do not need a speedy connection , you can save money with a 56K baud modem connection. You can upgrade for a DSL connection for faster service. A cable modem will deliver even faster service than DSL.

Having tested 5 of the ISP listed in figure 39 over 7 years, the author can vouch for the fast speed of the cable modem and cable service over the 56 KB dial up service of most of the ISPs. All ISPs had web hosting services but at various levels of sophistication. For example, one of them, Earthlink, offers an FTP utility built into the software for the end user to put information into his web directory. Some offer DSL (Digital Services Line) service for a slightly higher price. It is faster than 56K but not as fast as cable modem. Some of the ISP did not have enough alternate phone numbers in nearby small towns and cities and phone lines to call if the one you were dialing was busy. Caveta emptort when buying these ISP services because I had a variety of support satisfaction levels for each of the companies. One even double charged me for the service for 6 months without my consent or any reason (except profit motive). You may also setup a wireless LAN on a router in your home network. Verizon does this for free when you purchase the bundled package of services for fiber optic FIOS TV/voice/internet.

Modern Communications Systems

Figure 40 shows the hardware required for communications using an ISP. It is estimated that in today's world the children and adults need an internet connection to keep up with business at the new speeds enabled by internet.

Figure 40. Hardware Requirements

Video Display – high quality plasma screen or VDT

IBM PC – Pentium III and above with RS 232 ports and USB ports

Hard Drive – 2GB or greater

CDROM drive – Read / Write or Read Only

Modem – 56K Modem or Cable Modem (internal or external)

Twisted Pair Wiring or Coax Cable

(the above are one time costs and can be bought on credit and taken off taxes)

Wireless 802.11 Router

ISP Service or Cable Service – monthly cost

Figure 41 shows the suggested software for a setup at your home system for point-to-point communications.

Modern Communications Systems

Figure 41. Suggested Software

Windows XP Home

McAffee Virus Scan

Internet Explorer

Turbotax

Microsoft Office XP

Microsoft Publisher

Adobe Acrobat Reader

Winzip Compression Utility

ISP Installed Software

NETTOOLS5 from TUCOWS freeware

In the end you will do what you can afford to do on your budget. If you are wise, you can recoup much of the cost of your system from your annual taxes for business pruposes only. It is a good reason to build a home office where you can do your business at home.

Modern Communications Systems

11. Local Area Networks

Local area networks are defined as those which are contained in one building or room. They start when the computers users realize they would save resources by sharing printers, plotters, storage, work stations, and CDROM towers. They have a certain basic network topology based on the wiring schema and are this is the way to manage the installation of the network as a COTR (Contracting Officer's Technical Representative). The COTR budgets for the network system and then ensures implementation with the LAN Administrator. The LAN Administrator recruits and trains willing assistants and then puts them on the front line of managing the installation and maintenance of the project. He holds weekly meetings and asks for results. He ensures the assistants can train end users in all applications software as well as install applications software and troubleshoot operating systems problems. They, in turn, teach the network applications to end users and report back to the LAN Administrator. They monitor people online with a spy utility to see if they are having typing problems or other problems related to human factors. The network builds require team play from everyone involved. The LAN Administrator's task is as a coordinator and leader of the network fitness program. It is his sole job to keep the local area network up to speed with new hardware and software and applications. LANs became popular in the late 1980's and early 1990's and in Washington a lot of agencies invested a lot of money in downsized LANs that depended solely on servers and no mainframe connections. This was a change from the past strategy of buying mainframe technologies at data centers. The change allowed the PC to become a desktop mainframe and used by power users and regular users alike and it allowed decentralization of the IT function. The main problem here is that you now have unqualified, non-mathematical, untrained staff in IT positions who think they know all about IT because they can run software applications on a personal computer. These people are dangerous and the LAN administrator knows who they are if he has had experience on mainframe systems programming and mini computer networks. The LAN was written about by James Martin who is the greatest single author in computing and he showed the world how LANs would change the office environment in the 1980's. His vision became a reality for many government, private, and academic organizations. Technically, there are several; types of LAN architecture and all the other larger LANs are built using the LAN as a group of linked computers.

Ethernet topology – a bus topology where the LAN is connected and communicates using a multi-line bus drop and CDMA-CD as broadcast to the network addresses on the drop through the NIC card on each PC.

Figure 42. Ethernet

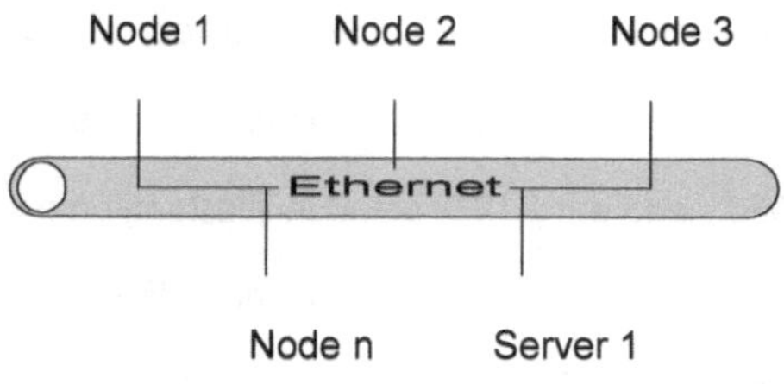

96

Star Topology – the older topology of many mainframe and mini-computers and the wiring actually look like a star configuration.

Figure 43. Star

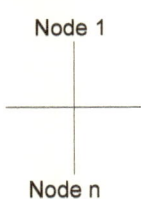

Hub Topology – similar to the star topology with all the network nodes surrounding the center hub where the main CPU is located. There may be more than one hub in the network where a lot of traffic comes through that machine. This is analogous to the airline industry hub systems of airports that carriers fly daily in America.

Figure 44. Hub

Ring Topology – the circular set of nodes that communicates by a token one message at a time and then forwards the message to the next computer node in the network interrogates the token and either puts the message in the token or forwards it to the next machine address in the ring. If the token is captured and the address matches the computer then the message is processed by that computer.

Figure 45. Ring

Wireless LAN 802.11 - Wireless routers allow a server to communicate with other clients in your home or office network. Any PC can be a server. The router can be configured with software that monitors the LAN connections to laptops and other PCs with network 802.11 cards in them. A WEP address then is needed on the client PCs for secure connection to the wireless router and server.

LANs have the following equipment requirements:

Figure 46. LAN Hardware

Routers (also called switches)

Bridges

Gateways

Hubs

Servers

Clients

Modern Communications Systems

Transmission Media (Fiber Optic, Coax, Twisted Pair)

NIC cards

Wireless Routers

Encryptors

Modems/Cable Modems

Protocol Converters

Figure 47. LAN Software

NAT – Network Address Tables

Protocol Stacks

RPC Programs

TCP/IP Stacks

OSI Model Standards

X.25

IPX/SPX

HDLC

POP3 Email Servers

SNMP

STMP

DIGITAL DECNET

ORACLE SQLNET

IBM TSO

Figure 48 shows the 20 activities that an effective LAN administrator needs to know and be able to implement. Note that applications programming, database programming, and systems analysis are not on the list although these are helpful to the LAN Administrator's background. Other people on the LAN have these functions and can assist the LAN administrator when he needs them. He also has to be able to delegate any of his tasks effectively and be a fast learner. He has to be able to master the network operating systems and integrate the workstations and client products on the LAN so that all users can bootstrap flawlessly when they need to.

Figure 48. LAN Administrator Functions

1. Network Security and Monitoring
2. Software and Hardware Acquisition
3. Network Configuration and Topology Planning
4. Network and Applications Contractor Management
5. Network Script Writing
6. Network Server Backups and Recovery
7. Network User Growth
8. Network Physical Space Growth (GB)
9. Server Management (Database, Communication, Tape Archive, CDROM, Email, Applications, Various OS)
10. Communications Server Management

11. Dial Up Communications Mgt

12. Special Mobile Equipment Management

13. Client and Workstation Systems Administration

14. Disaster Recovery and Contingency Planning

15. Preventive Maintenance

16. Cost Accountability

17. Team Leadership

18. Teaching Junior Staff

19. Reporting Projects Progress to Management

20. Communications with Network Users

It also helps to know the following standards, laws, and regulations for government network management:

Figure 49. Standards, Laws, and Regulations

NIST Standards

ANSI Standards

IEEE Standards

FCC Regulations

Procurement Laws

Personnel Laws & Regs

Modern Communications Systems

12. Wide Area Networks

Wide area networks are also campus network systems. They span several buildings in the same local area. They have the same components as the smaller LANs and are built with considerations for portals into larger network systems with gateways and bridges to those systems. Wide area networks are built with several LANs. They include multiple buildings. Today, several buildings that are smart buildings with pre-wired wall connectors for a network can help ease the installation process of a computer network. The building is pre-wired as an Ethernet or token ring network and the PC's are added as clients and a location for the repeaters, hubs, and servers is hidden from general access by the staff. Examples of a wide are network include a college campus, a government compound, an agency set of buildings on a military base commonly known as tenant commands, and a large single or multi-purpose facility. A countywide network for a public school system would be a wide area network. Most of the wide area networks can also communicate using TCP/IP and long haul communication such as Internet and email applications. Open systems communication really is about the ability to transmit outside the local area or wide area.

Figure 50 shows what the critical skills for an MVS systems programmer are for an IBM mainframe computer. This is the type of computer that can be used as a hub of a wide area network.

Figure 50. Critical Skills of an MVS Systems Programmer

1) MVS Utilities

2) MVS Security

3) MVS System Boot and Shutdown Sequence

4) MVS Scheduler – Batch and Interactive

5) TSO & CLISTS

6) Assembler Language and Macros

7) TCAM, VTAM telecomm methods

8) VSAM, ISAM, SAM access methods

9) IBM 2400 Tape Drives

10) IBM 3774 Controllers

11) IBM Terminals

12) IBM DASD Drives and Disks

13) Third Party Utilities for Backup And Recovery

14) Offsite Contingency Planning

15) Job Resource Accounting

16) Operations Procedures

17) Protocols

18) Tape Library

Modern Communications Systems

IBM SNA

    IBM Systems Network Architecture (SNA) was developed at the same time as the OSI model and is very similar in mission and construct. SNA has 7 layers for a mainframe communications protocol. The lower 3 layers of the SNA model are X.25 which any IBM computer can communicate (physical, network, transport layers). The upper layers define the HDLC protocol. Computers like the IBM 3090, 370, 360 series communicate using this architecture. IBM PC's generally use the OSI model and SNA when connected to a mainframe computer. The IBM operating systems is called MVS/XA and links the components (terminals) to the computer using VTAM (Virtual Telecommunications Access Method). IBM MVS/XA is a16 bit assembly language architecture machine. An SNA gateway controller connects the mainframe to the X.25 network. IBM 3074 controllers are the older dumb terminal controller in an IBM network of terminals.

Modern Communications Systems

13. Global Area Networks

Global Area Networks are those that span the globe and can usually rely on satellites technology and speedy transmission. Internet is a global area network. The new internet is a global internet that is improved. Signals can be sent and converted from any media to satellite channels and back again. Satellites make global networks better and faster than any other media of transmission. Today, we are watching the space race for communications. Space junk is everywhere in the earth's atmosphere. NASA even has a website to track all space objects that are man made. NORAD also tracks this and used in the Columbia Accident Investigation to track the piece of Space Shuttle that came off the launch vehicle. The extra piece of space junk from the launch vehicle orbited the earth and should have been a harbinger of things to come. I think from now on when an extra piece of space junk appears after a launch NASA will take more appropriate precautions to investigate it and take proper safety precautions. The shuttle puts many communications satellites into service at lowest taxpayer cost. The companies pay a fee for delivery of the satellites but it does not cover the majority of the cost of the launch. The Iridium network of satellites is used for secure satellite telephone. AFCEA Signal magazine sells Iridium phones for a quite expensive price to the normal hobbyist. Forouzan discusses the Iridium network in his book[11]. It has 77 satellites and was started in 1990 by Motorola. The name Iridium comes from the 77th chemical element

---

[11] Forouzan, Data Communications and Networks, McGraw-Hill, pp. 426.

on the revised periodic chart. The satellites are in a LEO orbit at an at altitude of 750 km above the earth. Each satellite has 48 spot beams on various land masses on the earth and the total system can have up to 3168 beams. These focus in on where the repeated signal can be transmitted and received by ground station parabolic dishes and converted to cable signals and other transmission media transparent to the sender and receiver.

Another NASA website tracks the actual orbits of the Satellites by name. This website really shows how many satellites are out there except for classified ones. The next version of internet called "Internet in the Sky" will be satellite based and is funded by Microsoft and Teledesic. The Teledesic network has 288 satellites in 12 LEO orbits at an altitude of 1350 km. The project is expected to be completed in 2005[12]. This has also been called "Internet 2". The advantage of satellite communications is speed and flexibility. The broadband channel provides a most convenient method of sending signals around the globe. Since a satellite is only a repeater in the sky, they move long haul data at a very fast rate in packets. Every nanosecond an electronic signal travels one foot according to the late Admiral Grace Hopper. A few million nanoseconds stacked end-to-end and you have a few second delay time to uplink and downlink. The speed of Internet 2 will be great until the system becomes saturated with messages. There will be no peak and non-peak times of transmission in all probability based on global simultaneous usage. If the system becomes very

---

[12] Forouzan, Data Communications and Networks, McGraw-Hill, pp. 427.

saturated, there can only be a finite number of satellites around the earth and the solution may not be more.

Interstellar space travel has affected global networks by the transfer of technology. Solar sails on satellites are expected to propel them into the far reaches of the universe where they could not go before. This means that global networks will quickly become universal networks certainly to the end of our solar system and well into others. Funding does not allow manned space flight at the rates we saw in the 1960's with other priorities here on earth. Unmanned missions as sensors to return data from the far reaches of the galaxy are a good solution and this strategy is being used at NASA. Mars missions and other deep planetary missions will beam back information in real time to our systems here on earth (if a 12 minute delay is real time). Creating a deep space network of listening posts and active research vehicles is an alternative to manned space flights we have taken. Our global networks should improve as we further penetrate deep space and understand other planets better. The SETI (Search for Extra Terrestrial Initiative) program allows us to participate in listening into the universe for a response to our transmissions to find other life through SETI@home connections[13]. This can be monitored at the home using a desktop computer by any citizen. We are in a truly remarkable scientific time when the average person has access to so much through his personal computer and modem that he can engage the universe at large without any extensive expertise.

---

[13] Panko, Business Data Communications Networks and Telecommunications, Prentice–Hall, pp. 377.

Modern Communications Systems

Most global networks are transparent to the sender of the messages. He doesn't care how the message gets to the destination, but rather that it does get to the destination using the communications models and networks of today.

Through ignorance Hollywood has both helped and hurt the cause of global communications networks. Movies such as The Terminator, The Matrix, and others focus on the negative impacts of technology and give the public the wrong idea about most real technology in orbit. Although the DOD does run a missile defense system, it is under total control of the president. It is the communications satellites and sensors that tell us what other countries are doing. These movies are designed to sell tickets and be entertaining at the expense of telling the truth. Since the 1970's with Hollywood disasters movies, they have focused on the negatives aspects of high technologies and disperse half truths to the public, possibly scaring the layman. Only through education in the sciences and engineering can one appreciate how much TV and the movies have shaped the public view of high technologies. This has been detrimental to the truth about satellite communications and other valuable space research projects. If we cancel all space research, we get on our own course towards ignorance of science and future discoveries.

Modern Communications Systems

14. Satellite Networks

NASA does technology right. The NASA website has media kits that explain every space probe we are using to explore the galaxy. NASA had long haul data communications down in the 1960's and transferred many ideas to the general public such as CDROMs for recording large streams of telecommunicated data. Video streams were transmitted from the moon in 1969 in bands or strips and put together here on earth. The museum at Goddard Space Flight Center has evidence of these early applications of this technology using gold plated CDROMS for high integrity. Our ground stations listen in for extra terrestrial communications from other planets as we communicate our secret code of life to the stars and beyond, hoping for a good answer for the universe. Space telescopes (Hubble) let us see into the corners of the universe. The space station provides a point of congregation for military and communications satellite maintenance. We have more satellites and space junk out in space now than ever before. NORAD tracks all the space junk. It's a wonder the shuttle doesn't have to alter it's path to get through the belt of space junk and LEO (Low Earth Orbit) and MEO (Medium Earth Orbit) satellites. Thank God for coordinated systems in use during shuttle trajectory analysis and launches. Many of the major players in the communications satellite industry are private sector companies with records of successful flight in space such as Hughes, GE, and COMSAT. The first private launching of space craft took place on June 20, 2004. Dr. Edwin Aldrin predicted this in 1998 in a speech at

Modern Communications Systems

American University for adventure space travel only. Older projects are available on the internet and include CORONA, TELSTAR, SPOT Image, LANDSAT, and others. NOAA runs some weather satellites that beam back earth weather patterns to us. We have GPS technology navigation satellites directing all types of devices here on earth. The Iridium network of satellites is a type of communications satellite radio that is very expensive and covers the earth with many satellites in LEO. The satellite frequencies are listed in the chapter on the frequency spectrum.

Below is a listing of satellite networks and network providers that carry educationally relevant material. Some networks may have a charge associated with their service.

Table 3. Educational Satellite Networks

CODE	NETWORK	SATELLITE	RECEIVER*
ATV	ACHIEVEMENT TELEVISION	Galaxy 3 Ch. 21	C
CPB	ANNENBERG/CPB CHANNEL	GE-3 Ch. 51	G
CSPN	C-SPAN	Satcom 3 Ch. 7	C
CSPN	C-SPAN 2	Satcom 4 Ch. 19	C
NASA	NATIONAL AERONAUTICS & SPACE ADMINISTRATION	GE-2, Ch. 9	C
PBS	PUBLIC BROADCASTING SERVICE	Telstar 402 Ch. 18 GE-3 Ch. 500-513	C G
SEA	SEAWORLD-SHAMU-TV	Telstar 4 Ch. 9	C
SERC	SATELLITE EDUCATIONAL RESOURCES CONSORTIUM	Telstar-402 Ch. 11 Ch. 323-331	C G
SITC	HEB SATELLITE IN THE CLASSROOM	Telstar 5 Ch. 23 Telstar 4 Ch. 513	C G
TEAMS	TEAMS Distance Learning	SBS6 Ch. 3,12	C

| CUSDOE | US-DOE SATELLITE TOWNHALL MEETINGS | Galaxy 9 Ch. 2 SBS6 Ch. 12 | C C |

Source: www.klvx.org

Teledesic and Internet in the Sky[14]

Teledesic is building a global, broadband Internet-in-the-Sky® network. Using advanced satellite technology, Teledesic and its partners are creating a communications network that will enable cost-effective access to telecommunications services such as computer networking, broadband Internet access, interactive multimedia and high-quality voice. The Teledesic Network is designed to meet the broadband needs of government, business, non-profit organizations, and individuals on a global basis. Teledesic is a private company based in Bellevue, Washington, a suburb of Seattle. It is not accidentally close to Microsoft in nearby Redmond, Washington.

Table 4. Teledesic Timeline

1990 Company founded

1994 Initial system design completed; Federal Communications Commission application filed

1997 FCC license granted; World Radio Conference designates necessary international spectrum for service

1999 Teledesic signs major launch contract with Lockheed Martin

2002 Teledesic signs contract with Italian satellite manufacturer Alenia Spazio SpA to build two satellites for Teledesic's global, broadband communications network

2005 Service targeted to begin

---

[14] Internet webpage http://www.teledesic.com accessed on 8 April 2004

Modern Communications Systems

Iridium Satellite Phone Network

The Iridium Satellite Phone network is a network of 66 satellites circling the earth for telephonic conversations by government personnel. Bill Clinton used this system. The primary customer is DOD and other government customers. This system is primarily for secured telephone class globally. The National Cryptologic museum has an Iridium telephone used by the presidents on display.

Figure 51. Iridium 66 Satellite coverage

NASA's website is very upbeat and designed to educate the public on good science. As such we need to understand more about NASA's role in

Modern Communications Systems

scientific research and grants to discover new sciences. Figure 32 is a sample of NASA's website.

Figure 52. NASA Website

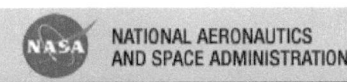

National Aeronautics and Space Administration · HUMAN SPACEFLIGHT

+ SHUTTLE + STATION + REALTIME DATA + NEWS + GALLERY + QUESTIONS + HISTORY + INFO + SITEMAP + SEARCH

 Space Shuttle

NASA Names New STS-114 Crewmembers

 NASA announced the addition of three new crewmembers for the next Space Shuttle mission, STS-114. Meanwhile, the Agency continues its return to flight efforts for the Space Shuttle fleet. STS-114 preflight images.

▪Check out the Return to Flight Reference Page and the STS-107 Investigation Reference Page.
▪View the CAIB Final Report: Low res (10 Mb PDF) and medium res (28 Mb PDF) versions.

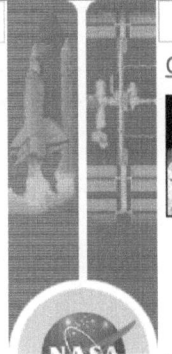

 Space Station

Crew Operates Robot Arm

 The Expedition 8 crew's Thursday aboard the International Space Station was filled with robotic arm operations and medical exams.

▪Expedition 8 Press Kit
▪Crews Complete Third Year on ISS
▪Read Expedition 7 Astronaut Ed Lu's last letter from the Space Station.

 Behind the Scenes

Behind the Scenes of Human Space Flight

WEBLAUNCHPAD

 Space News

NASA Names Crew for STS-121

Modern Communications Systems

Human space flight starts on the ground, where thousands of NASA employees, contractors and industry partners work together to send humans safely into space. Meet the people who make it all happen, and visit the unique facilities where they work in Behind the Scenes.

Check SkyWatch for space station sighting opportunities in your city.

ISS Status Report No. 63

ISS Crew and Students Are 'Partners In Flight'

Columbia Anthem Gets Grammy Nod

NASA Scientist Trains Astronauts to be 'Earth-Smart'

Return to Flight Implementation Plan Revision 1.1

NASA Names Next International Space Station Crew

Implementation Plan for International Space Station Continuing Flight (828 Kb PDF)

Expedition 8 Mission Overview (PDF 868 Kb)

Where is the Station?

NASA tabbed veteran Astronaut Steven Lindsey to command STS-121, a Space Shuttle Discovery mission recently added to the flight manifest. In addition to Lindsey, the STS-121 crew will consist of veteran Astronauts Mark Kelly and Carlos Noriega and first-time space traveler Michael Fossum. Other crewmembers will be named later.

.Check out the new Apollo interactive.

15. RPC Programming

Remote Procedure Calls (RPC) are a way programming that takes advantage of tunneling under the security of a network connection. The best book on this topic is one from O'Reilly called "RPC Programming". You'll need access to two computers that are connected so that you can bind a program together with the protocols necessary to make the connection through socket connections setup by the protocol on both ends of the point-to-point connection. The connection stays active during the execution of the program. I taught a course once called Distributed Systems and one student wrote a nice RPC program as his project for the term. I learned a lot from him about how to write the program using protocols for point to point connectors. Discovering RPC programming is like discovering a great secret of communications programming since most LAN administrators won't teach you how to write these types of programs. In fact, that student's boss had problems with him learning how to write RPC programs in a secure military computer environment. The reason: RPC programming obviates the security of the connections between computers. Once you learn RPC programming, you can take control of any computer that doesn't defend against an RPC socket connection. That fact alone makes RPC programming one of those items you don't teach everybody in a data communication or networking class. You can bet the boys at NSA and SANs Institute know about it though.

Modern Communications Systems

Figure 53 shows the RPC programming process. It is critical and difficult to test if you do not have two computers setup. This is because it connects the two ends of the protocol together by binding them and then manages the data connection. These programs can also be difficult to debug. We have not shown any source code for UNIX RPC programming here but refer to O'Reilly Power Programming in RPC for the best source code listings. The interesting fact about compiling RPC programs is that they take two computers to compile or bind together the protocol and can be rather easy once you understand the source coding scheme.

Figure 53. RPC process diagram[15] overview

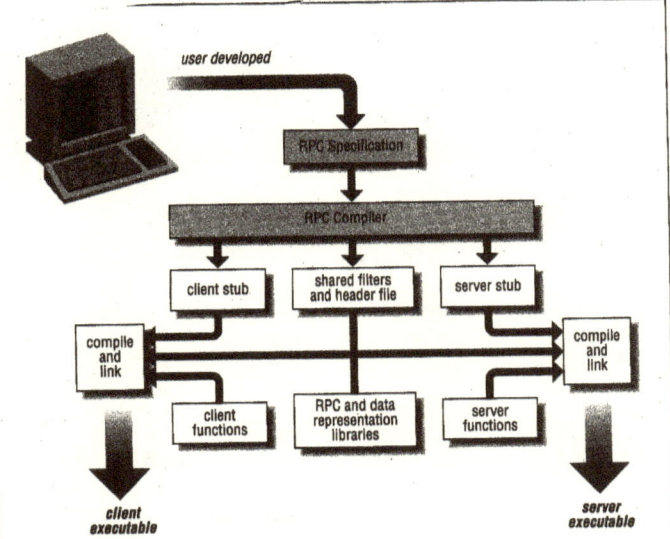

The development of an application that makes use of lower-level RPC function calls requiring an additional step: the compilation of a protocol or RPC specification. With a protocol compiler you do not have to perform RPC communications in your client and server code.

The ONC RPC protocol compiler, RPCGEN, produces client and server stubs that use lower-level RPC calls. RPCGEN allows more flexibility in building the client and server with only a modest increase in programming complexity.

---

[15] O'Reilly, Power Programming with RPC

16. Communications Security

The literature abounds with new books on computer security like CyberWar 2.0, the SANS Institute webpage, guides on Intrusion Detection, and new software products from companies like TripWire, R.S. Carsons (Saint), McAffee Viruscan, and others. The ZDNet website has free shareware for computer security that has grown over the years and allows normal people to use various security products. Many countries now export their computer security products (encryption) to the United States especially United Kingdom and Germany. A simple book on a communications technology like SNMP can enlighten a normal programmer to the possibilities available in communications security. The times of computer security being only a function of many function is long past. Today, the CISSP certification test and CISO (Chief Information Security Officers) are making the specialty a real interesting one for young people in computing. Several colleges and universities have recreated their computer departments to include a computer security track to keep up with the demand for highly qualified computer security people in the public organizations.

Computer and communications security is best implemented in a layered approach like the onion approach of James Martin[16]. The 9 layers of the security onion from the outermost to the innermost layer are:

---

[16] Martin, James, Computer Security and Privacy, Prentice-Hall, 1982.

Modern Communications Systems

Figure 54. Nine Layers of Computer Security Onion

1. Political Security

2. Legislative and Policy Security

3. Administrative Procedures Security

4. Personnel Security

5. Environmental Security

6. Physical Security

7. Hardware Security

8. Firmware Security

9. Password Security

Communications security also includes vulnerability analysis and testing. Virus testing and debugging is also a key feature of successful communications security plans. An installation needs a communications security plan that can reduce risks and mitigate security breeches in an IT Contingency Plan. The threats matrix is the way to communicate this plan and implement the elements of the plan. Another name for this plan is the Business Continuity Plan or the Disaster Recovery Plan. The idea is to have written down what the threats are to that computer/communications network environment and create solutions to the problems to mitigate the risks at least cost to the network owner. In government the agency is the owner of the network even when they are private networks.

Modern Communications Systems

The types of systems that have security are voice, IP, network, cellular, and server hosts. These systems all are vulnerable to attack and security breeches. The SANS Institute is the authority that NSA trusts and they are the advisor to the president of the United States on Cyber security. The FBI and DOJ have recently entered the computer security field and are also major player in stopping all white collar crime. Every major military branch has their own CERT (Computer Emergency Response Team) group and we have one nationally at Carnegie Mellon University as created by Congress.

Good security is something that we must afford and ensure that we take action before any crimes are committed. The return on security (ROS) is that we are able to process our data requirements and communications requirements more securely with more privacy and efficiency than ever before in the history of electronics with greater returns for our common good and national security. This is a concept suggested to government computer people from NSA policies on communications security.

Modern Communications Systems

17. Communications Associations

This section will allow you to understand and join some of the major

technology associations that deal with telecommunication and data

communications on a regular basis. The primary associations are IEEE, ACM,

ICCP, AITP, SANS Institute, ICCP, and AFCEA. A more local place to get

started is your local Radio Shack who has a lot of the commercial systems you

can build and operate as a hobby. Here are there full names and addresses:

Figure 55. Electronic Communications Associations

IEEE – Institute for Electrical and Electronics Engineers
     445 Hoes Lane
     PO Box 1331
     Piscataway, NJ   088555-1331 USA

ACM – Association of Computing Machinery
     PO Box 11315
     New York, NY   10286-1315 USA

AFCEA – Armed Forces Communications and Electronics Association
     4400 Fair Lakes Court
     Fairfax, Virginia 22033-3899 USA

AITP – Association of Information Technology Professionals
     Previously Data Processing Management Association (DPMA)

ESI International – ESI International
     4301 Fairfax Drive, Suite 800
     Arlington, Virginia 22203 USA

SANS Institute – Systems Administration and Network Security Institute
     8120 Woodmont Avenue Suite 205
     Bethesda, Maryland 20814

ICCP – Institute of Certified Computer Professionals
     2350 East Devon Ave. Suite 115

Modern Communications Systems

Chicago, IL. 60018-4610

ISACA – Information Systems Audit and Control Association
3701 Algonquin Road Suite 1010
Rolling Meadows, Ill. 60008 USA

Federal Communications Commission
Independent US government agency directly responsible to Congress, regulates interstate and international communications by radio, television, wire, satellite and cable.

IEEE Communications Society
Promotes developments toward meeting new market demands in systems, products, and technologies such as personal communications services, multimedia communications systems, enterprise networks, and optical communications systems.

International Engineering Consortium
The International Engineering Consortium (IEC) conducts a broad range of university and industry cooperative programs consisting of educational forums and workshops, research studies, publications, Web education, and management services.

International Telecommunications Union (ITU)
An international organization within which governments and the private sector coordinate global telecom networks and services. Headquartered in Geneva, Switzerland.

Mobile Satellite Users Association
A non-profit association to promote the interests of users of mobile satellite communications worldwide. It fosters effective communication among Mobile Satellite Services (MSS) users, suppliers of equipment and services, operators of the satellites....

NAB - National Association of Broadcasters
Based in Washington DC, NAB represents the radio and television industries before Congress, the FCC and federal agencies, the courts, and internationally. NAB provides its resources to support members, broadcasters at-large, and through ongoing...

National Telecommunications and Information Administration
An agency of the U.S. Department of Commerce, NTIA is the Executive Branch's principal voice on domestic and international telecommunications and information technology issues.

Modern Communications Systems

Satellite Communications Systems and Technology Group
A study covering emerging systems concepts, and applications. They studied the international status of satellite communications systems and technology. These included major manufacturers, government organizations, service providers, and associations.

Satellite Industry Association
Represents the U.S. commercial satellite industry. Member companies are the leading satellite service providers, satellite manufacturers, launch services companies and ground equipment suppliers in America.

The Center for Satellite and Hybrid Communication Networks (CSHCN)
The goal of the center is to become a leader in the crucial convergence of satellite and terrestrial communications technologies.

The International Institute for Communication and Development (IICD)
IICD acts as an independent broker between countries in development and the stakeholders that drive the international market of ICTs.

UMTS - Universal Mobile Telecommunications System
The UMTS Forum works as a catalyst with other specialist organizations to examine issues such as technical standards, spectrum, market demand, business opportunities, terminal equipment circulation and convergence between the mobile communication...

US Department of Commerce Office of Telecommunications
Provides information regarding various forms of transportation such as cable, cellular and satellite.

VITA: Volunteers In Technical Assistance
VITA's offers information dissemination techniques such a communications technologies of digital radio networks and a low-earth orbiting satellite system, VITAsat.

Some of these associations have certification exams you can take to improve

your professionalism. Most allow you to join as a member with certain

qualifications and basic education requirements in the areas of concern.

18. Communications Equipment Manufacturers

Communication equipment manufacturing is an important trade to our national economy and trade surplus with other nations. The United States is a leader in communication equipment and many subparts are manufactured overseas where labor rates are lower for subcomponent assembly. Table 5 shows some of the communications industry links to equipment manufacturers, computer, satellite, terrestrial and TVRO.

Table 5. Communications Industry Links to: Equipment Manufacturers, Computer, Satellite, Terrestrial and TVRO	
AcuSat	AcuSat satellite software - commercial satellite software company
ADL	Antenna Downlink, Inc.
ARC	Antennas America, Inc.
Advanced Technical Materials, Inc.	Manufactures, designs and stocks microwave, coaxial and rf components
Advantech	PC based monitoring and control systems
Advent Commications LTD	Manufacturer / supplier of products, subsystems and systems to the satellite broadcast and telecommunications industry.
Alcatel	Network communication provider and cable manufacturer
All Mobile Video	Manufacture, rental and sales of a broad range of communication equipment and vehicles.
American Communication Systems	Wireless communications equipment and services provider
Amphenol Corporation	Manufacturer of connectors and cable assemblies

Modern Communications Systems

Andrew Corporation	**Andrew makes a broad range of products for the communications industry including connectors, cables, towers and antennas.**
Antek Systems, LLC	Dedicated to providing superior products and support services for earth stations around the world.
Avdata	**Design and management of digital communication networks**
Boeing Space Systems	Boeing Aerospace Corporation
BT Broadcast Services	European supplier of broadcast solutions providing a comprehensive range of terrestrial and satellite services for international television and radio broadcasters.
CableLabs	**Cable Television Laboratories, Inc.**
California Amplifier	**Manufacturer of satellite, wireless cable, voice and data components**
Channel Master	Manufacturer of satellite and terrestrial communications equipment.
Chapparral	**A source for satellite antennas and receivers**
CME	**Italian company manufacturing Universal Remote Controls**
Commercial Satellite Systems, Inc.	**Provider of satellite communication systems, primarily VSAT**
Comstream/Radyne	**A complete line of earth station and microwave equipment and systems**
Comtech Antenna Systems, Inc.	**Satellite Antennas, Mobile Antenna Systems and C, X and Ku Band Flyaway Antennas**
Comtech Systems, Inc.	**Digital Troposcatter Systems, Multiplex Systems and High Performance Microwave Radios**
CPII/Communication and Power Industries	**Provider of Vacuum Electron devices, subsystems and amplifiers**
Cross Technologies	Designs and manufactures electronics for the satellite communications, broadcast, cable television (CATV), microwave and general communications industries.

Modern Communications Systems

DNE Technologies, Inc.	**Manufacturer of datacommunication multiplexers**
EADS	**Transnational Aerospace Company – Conglomerate**
Eagle Communications	**Plastic Injection Molding, Powder Coating, Antenna Manufacturing and Installation**
Galaxis	**European based manufacturer of a wide variety of TVRO equipment and a dish of a different color.**
Gardiner Communications	**Designer & Manufacturer of Satellite TVRO Systems & Components**
Gavilan Communications	Digital MPEG-2 DVB and analog satellite receiving equipment and accessories
General Instruments	**Manufacturer of a broad range of equipment for communication applications**
Gourmet . . . Entertaining	**Provider of quality tools for satellite dish installation**
Hughes Communications	**Part of the Hughes Aerospace/Communications family**
Kaul-Tronics, Inc. **also known as KTI**	**Producers of mesh antennas and offset antennas**
Leitch	**Producers of a large line of communication and production electronics.**
Lockheed/Martin	**Producers of aerospace systems, spacecraft, and electronics**
M/A-Com	M/A-COM has the world's broadest offering of RF, microwave and millimeter wave devices, components and subsystems.
Microtech, Inc	**Flexible and rigid waveguide assemblies and passive microwave components**
Microwave Filter Company	**MFC designs, manufactures and sells passive electronic filters for radio and microwave frequencies**
Miteq	**Satellite communication products - components and assemblies Now owners of**

Modern Communications Systems

	MCL, Inc.
Motorola	Motorola Electronics division
Norsat	Norsat designs, engineers and distributes products for use in the satellite wireless communications and cable television industries.
Orbitron International	Manufacturer and distributor of complete satellite systems
Paraclipse Corporation	Manufacturer of satellite antennas
Patriot Antenna Systems	Manufacturer of satellite antennas
Pico Products, Inc.	Manufacturer of high-performance electronics for the CATV industry
Pro-Brand International	Carrying a full line of satellite antenna hardware
Prodelin Corporation	A huge line of satellite antennas for varying applications
R.L.Drake, Co.	A wide range of communications and television products
Radyne/Comstream	Satellite and cable equipment - Tiernan digital compression technology
Real World Technology, Ltd.	The U.K's set-top box tuner specialists
SAMI	Superior Antenna Manufacturing, Inc.
SatMaster	SatMaster satellite software - commercial satellite software company
Scientific Atlanta	Sci. Atlanta offers a broad range of satellite eommunications technology
Superior Satellite Engineers, Inc.	Provider of Satellite Antenna Systems for Cable, Broadcast, Private and Wireless Television
Toshiba America	Manufacturer of a wide range of communication products
Uniphase Broadband Products	Specializes in development, manufacture

	and marketing of high-speed fiber optic communication products and test equipment.
Wegener Communications	**Advanced digital technologies**
Winegard Corporation	**Manufacturer of quality television reception products.**

Source: http://members.tripod.com/The_Uplinker/Links/equipment.html

America is not the only global player in the telecommunications game and Figure 56 shows the United Kingdom Data Communications Companies. Some of these companies may do business with American companies as well as the rest of the world on a regular basis. American companies often locate in the United Kingdom as they locate here. Since we share a common language, the business should be easy to work with these companies.

Figure 56. United Kingdom Data Communications Companies

Lan-Com International, Witney, Oxon, United Kingdom
Computer cable installation & computer network services. In addition computer installation local area network systems distributors. Also computer network equipment.
Email Website

Ring Communications (UK) Ltd, Ely, Cambridgeshire, United Kingdom
Provision of satellite tracking & location services for fleets, plant & machinery, & other mobile assets. Communication specialists for voice & video over IP, Intercom & professional Broadcast equipment

Jekyll Electronic Technology Ltd, Reading, United Kingdom
Electronic equipment designers & manufacturers

Datacom Technology, Nottingham, United Kingdom

LINKS Network Communications Ltd., Colchester, Essex, United Kingdom
Structured cabling, Cat 5e, Cat 6, fiber optics, network communications, computer cabling

CORNET TECHNOLOGY, Frankfurt am Main, Germany, United Kingdom
Cornet Technology designs, manufactures, markets and services vendor-independent hardware solutions for effectively switching, managing and

monitoring lines in data, voice and video communication. CORNET Technology's solutions replace traditional patch panels, automating a variety of time consuming, manual processes.

Radnet Ltd, Rochester, Kent, United Kingdom
Radnet is a dynamic and innovative provider of the full data networking life cycle, comprising network consultancy, design, documentation, installation, support and infrastructure procurement services.

3Com, Wokingham, Berks, United Kingdom
Data, voice & video communications

Acal Electronics Ltd, Fleet, Hampshire, United Kingdom
Electronic equipment & systems suppliers

Amplicon Liveline Ltd, Brighton, United Kingdom
Manufacturers of data communication systems; computers, industrial; computer systems (industrial automation) & data acquisition systems. Also converter, AC to DC; converters DC to DC; oscilloscope, power supply systems/unit power supply (switched mode) systems/unit; meters, panel, digital & signal conditioning equipment manufacturers

Automation Control Electronics, Wellingborough, Northants, United Kingdom
Computerised control systems

Avenue Communications Ltd, London, United Kingdom
Public Relations

Belgravium, Bradford, W. Yorkshire, United Kingdom
Radio data

Camtec Electronics Ltd, Leicester, United Kingdom
Data communication equipment

Case Communications, High Wycombe, Buckinghamshire, United Kingdom
Develop, manufacture & provide networking solutions (wire speed Linux routers & servers) also base repair, project management & pre & post sales support

Comlynx Communications Ltd, Birmingham, United Kingdom
Telecommunications, Cables & telephones

Comtec, Huntingdon, Cambs, United Kingdom
Telecommunication systems/equipment, cable locating equipment, data cable, fiber optic cable, fiber optic components/equipment, telecommunication test equipment, telephone cable, telephone, telephone line test equipment & television (cable) equipment distributors. Also manufacturers of headsets (communication)

Consultronics Europe Ltd, Eastleigh, Hants, United Kingdom
Telecommunication test equipment manufacturers

Datentechnik Intercom Ltd, Crowthorne, Berks, United Kingdom
Data communication systems manufacturers

Modern Communications Systems

Eloquence Ltd, Slough, United Kingdom
Data communication systems manufacturers

Enterasys Networks Ltd, Newbury, Berkshire, United Kingdom
Data communications

Globe Wireless, Liverpool, United Kingdom
Data communications

Hermes Datacommunications International Ltd, Shrewsbury, United Kingdom
Communication systems distributors

Icomm Structured Wiring Systems, Northampton, United Kingdom
Manufacture of structured cabling solutions

Kenwood Electronics UK Ltd, Watford, United Kingdom
Manufacturers of audio/hi-fi equipment/systems & two-way radio communication equipment. In addition mobile communication systems

KK Systems Ltd, Brighton, United Kingdom
Data communications equipment

KRONE (UK) Technique Ltd, Cheltenham, Glos, United Kingdom
Manufacturers of computer installation local area network systems enclosures/cabinets, fiber optics components/equipment/instruments & telecommunication components

Lancaster Communications Ltd, Lancaster, United Kingdom
Design & installation of computer network cabling systems

Metrodata Ltd, Egham, Surrey, United Kingdom
Data communication managed network services

Microcomms Ltd, Redruth, Cornwall, United Kingdom
Data cable communication systems manufacturers

Pact Equipment Ltd, Altrincham, Cheshire, United Kingdom
Data communications equipment manufacturers

Pierpoint Technology Ltd, Hassocks, W. Sussex, United Kingdom
Design and supply of electronic security systems

Pinnacle Communications Ltd, Rhyl, Clwyd, United Kingdom
Manufacturers fiber optic cables

Radius Computer Services Ltd (a division of Radius Plc), Feltham, Middx, United Kingdom Computer systems house

Remote Data Concepts Ltd, Nottingham, United Kingdom
Remote monitoring solution suppliers& manufacturers

Satelcom (UK) Ltd, Ascot, Berks, United Kingdom
Data communications equipment

Skyetronics (Technics) Ltd, Isle of Skye, United Kingdom
IT equipment

Modern Communications Systems

Team Simoco Ltd, Derby, United Kingdom
Analogue & digital radio for services

Telindus Ltd, Odiham, Hants, United Kingdom
Data communications equipment manufacturers

Trend Communications Ltd, High Wycombe, Bucks, United Kingdom
Communications test equipment suppliers

Vanguard Managed Solutions, Crawley, West Sussex, United Kingdom
Managed solutions provider

Jejo Ltd, Aylesbury, Buckinghamshire, United Kingdom

Source: http://www.kellysearch.com/gb-product-375.html

Ireland has some good data communications companies that American

companies can do business with easily due to the common language and culture.

The location of Ireland can be beneficial to some of the network nodes in Europe

to American based companies on this side of the Atlantic.

Figure 57. Irish Data Communications Companies

Bandwidth Telecommunications
Provider of commercial telecommunications services.
http://www.bandwidth.ie | Tuesday, March 05, 2002

Eircom
Details of the company and telecommunication services offered.
http://www.eircom.ie | Tuesday, March 05, 2002

Golden Pages
Searchable Irish telephone directories.
http://www.goldenpages.ie/ | Tuesday, March 05, 2002

HiberCall Ltd.
Details of telecommunication services offered.
http://www.hibercall.ie | Tuesday, March 05, 2002

Instant Communications Ltd.
Details of services including installation and maintenance of cable networks
(voice, data, video).
http://www.instantcomms.com | Tuesday, March 05, 2002

International Communications Users Group

information sharing for the business user, about network providers and suppliers in the telecommunications industry.
http://www.icug.ie | Tuesday, March 05, 2002

Ocean
Provides business telecommunications, subscription-free residential internet access and non-local residential telephone calls.
http://www.ocean.ie | Tuesday, March 05, 2002

Quinn Electrical Services
Design, installation and maintenance of power systems to the telecommunications industry.
http://www.qes.ie | Tuesday, March 05, 2002

Skycom
Providing software and email notification services to mail servers such as MS Exchange and Lotus Notes.
http://www.skycom.ie | Tuesday, March 05, 2002

Sord Data Systems
Info on local and wide area network solutions based on leading brand technologies.
http://www.sord.ie | Tuesday, March 05, 2002

Speedial Ltd.
Residential and business telecommunication services.
http://www.speedial.ie | Tuesday, March 05, 2002

Spirit
Provides residential routed telephone calls.
http://www.spiritelecom.ie/ | Tuesday, March 05, 2002

World-Link
Offers telecommunications services for home and business use.
http://www.nci.ie/telecoms/ | Tuesday, March 05, 2002

Source: http://www.gaire.com/~computers_data_communications.asp

The United Kingdom has some good satellite companies that one may wish to explore for services. Figure 58 shows these companies who are in the equipment business. This does not include launch vehicles to get the satellite airborne.

Figure 58. United Kingdom Satellite Communications Equipment

Autoleck, Welling, Kent, United Kingdom
Auto-electrical specialists. Full range of tracking systems, satellite
communication, car alarms and radios
Email Website

Hirschmann Electronics UK Ltd, Bedford, United Kingdom
Connectors & antennas
Email Website

Access Information Limited, Midhurst, West Sussex, United Kingdom
Supplier and installer of two way satellite broadband equipment. Supplier and
installer of the e2work GPRS interface, which allows remote users to reciprocate
their office desktop with full functionality in the field

Access Information Limited, Midhurst, West Sussex, United Kingdom
Suppliers and installers of the Hughes two way broadband over satellite.
Suppliers and installers of the e2work platform enabling personnel who spend a
lot of time away from their office a fast and secure connection to their corporate
LAN.

Aerials & Cable Equipment Ltd, Cheltenham, Gloucestershire, United Kingdom
Aerial & cable equipment distributors

Acal Electronics Ltd, Fleet, Hampshire, United Kingdom
Electronic equipment & systems suppliers

Advent Communication, Chesham, Bucks, United Kingdom
Satellite communications

Alphameric Solutions, Guildford, Surrey, United Kingdom
Satellite equipment manufacturers

A-N-D Group P.L.C., North Shields, Tyne & Wear, United Kingdom
Marine safety systems & marine electronics

Anritsu Ltd, Stevenage, Hertfordshire, United Kingdom
Elect components for satellite system manufacturers

Astrium Ltd, Stevenage, Herts, United Kingdom
Satellite & communications

Audio Visual Centre, London, United Kingdom
Mobile phones, Satellite, DVD

Broadcast Technology Ltd, Andover, Hants, United Kingdom
TV studio distribution equipment

Caprock UK Ltd, Aberdeen, United Kingdom
Satellite communication manufacturers/microwave systems

Channel 11 (UK) Ltd, Worthing, W. Sussex, United Kingdom
Satellite communications

Comtec, Huntingdon, Cambs, United Kingdom
Telecommunication systems/equipment, cable locating equipment, data cable, fiber optic cable, fiber optic components/equipment, telecommunication test equipment, telephone cable, telephone, telephone line test equipment & television (cable) equipment distributors. Also manufacturers of headsets (communication)

Data Marine Systems Ltd, Aberdeen, United Kingdom
Satellite television communications

Digital Sales Ltd, Chorley, Lancashire, United Kingdom
Satellite specialists

Double D Electronics Ltd, Gravesend, Kent, United Kingdom
Custom electronic systems

G C L, Newark, Notts, United Kingdom
Providing satellite broadband connectivity to businesses

Global Communications (UK) Ltd, Chelmsford, United Kingdom
Satellite communication equipment manufacturers

Hughes Network Systems Ltd, Milton Keynes, United Kingdom
Telecom systems

J.R.C (UK) Ltd, London, United Kingdom
Satellite communication equipment manufacturers

Livewire Digital Ltd, Epsom, Surrey, United Kingdom
Satellite communications

L-Teq Ltd, Frimley, Surrey, United Kingdom
Satellite communication systems

Merseyside Satellite Consultants Ltd, Liverpool, United Kingdom
Manufacture and wholesale of ariels

Mitec Telecom Ltd, Dunstable, Bedfordshire, United Kingdom
Microwave components manufacturers

Mitec Telecom Ltd, Dunstable, Bedfordshire, United Kingdom
Microwave networks & components

Next Destination Ltd, Salisbury, United Kingdom
Satellite communications

Nokia Home Communications Ltd, Swindon, United Kingdom
Digital satellite products

Paradigm Communications Systems Ltd, Alton, Hampshire, United Kingdom
Paradigm provides solutions for broadband terrestrial wireless & satellite communication networks

Peak Communications Ltd, Brighouse, W. Yorkshire, United Kingdom
Commercial satellite communications equipment suppliers & manufacturers

Precision Antennas Ltd, Stratford-Upon-Avon, Warwickshire, United Kingdom
Antennas, satellite dish & tower manufacturers

S P L - A C T, St. Ives, Cambs, United Kingdom
Satellite communications & defense

St. Bernard Composites Ltd, Farnborough, Hants, United Kingdom
Glass fiber moldings manufacturers

Scientific Atlanta Western Europe Ltd, Reading, United Kingdom
Global communications

Star Communication Consultants Ltd, Great Yarmouth, Norfolk, United Kingdom
Satellite communication systems

Strong (UK) Ltd, London, United Kingdom
Telecommunications systems distributors

Timestep Ltd, Dartmouth, Devon, United Kingdom
Satellite communications equipment manufacturers

Vega Group P.L.C., Welwyn Garden City, Herts, United Kingdom
Satellite telecommunication equipment

Vocality International Ltd, Godalming, Surrey, United Kingdom
Satellite multiplexes

L-teq, Camberley, Surrey, United Kingdom

Isis Electronics, Stroud, Gloucestershire, United Kingdom

TexoNet, Penicuik, Midlothian, United Kingdom

www.mesuk.net, Southampton, United Kingdom

Source: http://www.kellysearch.com/gb-product-80250.html

Cellular phone systems manufacturers are under great competition for the mobile computing and communications market. Figure 59 shows some of the companies who are very good manufacturers of cellular and paging equipment.

Figure 59. Manufacturers of Cellular and Paging Communications Systems

Modern Communications Systems

Advanced Mobile Solutions
International company that designs and manufactures compact power supplies for telecommunications equipment, computers, industrial controls and medical equipment.

Benefon
A specialized mobile phone manufacturer focusing on GSM and NMT technologies. Located in Salo, Finland

Bosch
Worldwide producer of mobile communications based in Stuttgart, Germany.

ELCOTEQ NETWORK OYJ (profile)
Elcoteq Network Oyj. The company manufactures electronics sub-assemblies and end products for the telecommunications industry. The company provides manufacturing services for digital mobile phones and their accessories.

Formosa Electronic Industries Inc.
Manufacturer of cellular and cordless phones, notebook PCs, and camcorder accessories. Based in Taiwan.

Globus Wireless (financials, news, profile)
Globus Wireless, Ltd. (formerly known as Globus Cellular, Ltd.) is involved in the research, design, manufacture, marketing and distribution of wireless communication products. The company announced a partnership with Auden Technology of Pa-Te c...

InfoSonics Corporation
Manufacturer of telecommunications accessories and distributor for numerous wireless phones. The primary customers are network operators, agents, resellers, dealers and retailers in the wireless communications market. Based in San Diego, CA.

Kyocera
CDMA phone manufacturer. Purchased the Qualcomm consumer phone business, including its phone and accessories inventory, manufacturing, and customer commitments in December, 1999.

LG Information & Communications
An integrated information/communication company which provides customers with advanced information/communication equipment including exchange systems, transmission devices and terminals.

Mitsubishi
Develops, manufactures and markets wireless integrated voice/data communications products and advanced mobile communications applications

and services.

Mobilex Cellular
A manufacturer of wireless accessory products, including Linear AC Power Supplies, Desktop Chargers, Switching-circuit Travel Chargers ("Switchers"), Vehicle Power Adaptors (VPAs), Wireless Connectivity Devices, and Hands Free Car Kits.

Motorola (financials, news, profile)
Motorola, Inc.. The principal activities of the Group are the provision of integrated communications solutions and embedded electronic solutions. These include software-enhanced wireless telephone, two-way radio, messaging and satellite communication....

NeoPoint
Produces "smart phones" and services aimed at optimizing the way wireless users reach people and information resources.

Nokia USA
Mobile phone supplier as well as a supplier of mobile and fixed telecom networks and services. Also creates solutions and products for fixed and wireless data communications. This site has Nokia mobile phone accessories available online.

Oi Electric CO., Ltd. (profile)
Oi Electric Co., Ltd. manufactures, sells, installs, and repairs various communications equipment including mobile communications equipment and measuring devices and parts. Mitsubishi Electric is the major shareholder of the company with 31.79% ...

Panasonic - Pagers
Offers numeric and alphanumeric pagers.

Promax Wireless Products
Supplier of after-market wireless accessories. Customers include: OEM, Carriers, Major Suppliers, Exporters, Wholesalers and Agents.

QUALCOMM (financials, news, profile)
Qualcomm Incorporated. The principal activities of the Group are the design, development, manufacture and market of digital wireless communications products and services; provision of license and receiving of royalty payment...

R.S. Communications, Inc.
Wholesaler of used cellular phones and refurbished cellular phones. Customers include: cellular carriers, agents, prepaid cellular phone dealers, and retailers.

Samsung

Producer of both single- and dual-band Code Division Multiple Access (CDMA) Personal Communications Service (PCS) handsets and CDMA cellular handsets in addition to other business and network solutions.

Shawcell Telecommunications Limited (profile)
ShawCell Telecommunications Limited provides telecommunications and financial services. The telecommunications business stems from two subsidiary companies namely Shaw Cellular and EuroPoint Cellular. ...

Sony Wireless
Division of Sony that manufactures and distributes single- and dual- band digital phones.

Unitech Industries
Designs, manufactures, and markets wireless cellular phones, computer and video accessories, battery packs and chargers, power supplies, and hands-free speakerphone kits.

Source:http://www.business.com/directory/telecommunications/wireless/personal _communications/cellular_and_paging/manufacturers/

Software companies play a major role in communications by enabling the machines we call computers to talk to one another using various programming languages to setup this talk. Sun's JAVA language was major improvement that affected Microsoft and other major manufacturer's in the marketplace. Novell basically cornered the market in LAN software or middleware for networks and is still going strong. Any good LAN manager needs to know these software sets to be effective on the job. Figure 60 shows some of the good software companies who provide software for communications and computer equipment. Some of the software companies are server software companies and some are distributed networking companies. All have some sort of software that runs on their hardware systems to connect computers together.

Figure 60. Software Companies

Apple

Borland

DBASE

Dell

Digital Electronic Corporation / Compaq

Gateway

Hewlett Packard

IBM

Microsoft

Oracle

Sun Microsystems

Sybase

Modern Communications Systems

19. Microwave Signals

Microwave telecommunications frequencies are high frequencies wide bandwidth delivery channels that are propagated from microwave dish tower to tower with the signal by line of sight.[17] The towers are located on high ground for best reception of signals. Microwaves were discovered by the radar industry as a heat signature during transmission. The frequency is from 1-30 Ghz. Between 1.85 GHz and 2.20 Ghz is used for many public services since 1992. The entire microwave bandwidth is used for voice, long and medium haul data, SCADA (Supervisory Control And Data Acquisition), and protective relaying. This implies some secret clearances required for work in this frequency range.

The signal works based on the parabolic equations we learned in trigonometry. The focus is the center of the parabola (dish) and the signal passes through the focus out through the air and can be intense or less intense depending on the signal strength. Since microwave was founded on radar principles, the signal has reflective properties like radar beams . Microwave signals also produce heat energy. This led to the discovery that one could pop popcorn with microwave beams. Microwaves have also been known to set off certain heart pacemakers.

---

[17] Fink & Beaty, Standard Handbook for Electrical Engineers, McGraw-Hill, pg 10-141.

The government has contracts to manage the towers and setup for signal processing in this medium. Most applications are secure on microwave stations in the state. The bandwidth and speed are excellent for most applications although there must be consideration for over utilization by non mission critical applications. Figure 61 shows a microwave tower installed by AT&T. There are also multipurpose towers which have microwave dishes on them.

Figure 61. Microwave Tower in Levan, Utah installed by AT&T

Source: www.drgibson.com

Modern Communications Systems

Figure 62. Microwave Tower in Michigan

Source: www.beaverisland.com

Figure 63. Microwave Tower in Glamis, California near Route 8

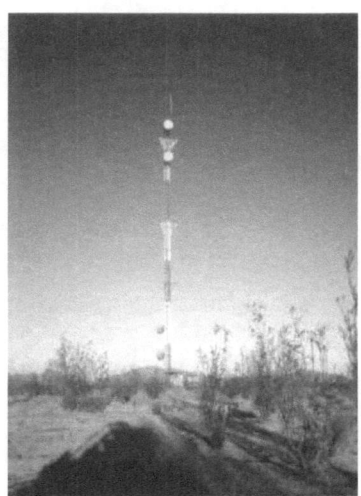

Source: www.glamisonline.org

Modern Communications Systems

20. Biometrics System Security

Biometrics are defined as a physical characteristic of a real person that are unique to that individual and uniquely identify him or her. Since physical security is one of the most assured types of security, Biometrics takes advantage of the mechanization of a physical characteristic such as eye retinal scanning, fingerprints, voice patterns, or facial characteristics. Law enforcement agencies through the FBI have access to criminal profiles of all these types and computer search algorithms can speed through them to select a person in very small time compared with older methods of investigation and criminal forensics. The Biometrics Consortium is an educational group sponsored by NIST who has testified to congress on the topic of Biometrics Security. Figure 64 shows the NIST Biometrics Consortium Topics as downloaded from the internet.

Figure 64. NIST Biometrics Consortium topics

Research:

Biometric Systems Laboratory (University of Bologna, Italy)

Center for Identification Technology Research (CITeR)

European Cooperation in the field of Scientific and Technical Research (COST)

Michigan State University Biometric Research Homepage

MIT Media Lab's Vision and Modeling Group

Modern Communications Systems

Ohio University Center for Automatic Identification

San Jose State University Biometric Research Center

Face:

Facial Analysis, (Preception Science Laboratory, University of California)

Facial Animation, (Preception Science Laboratory, University of California)

Face Recognition Technology (FERET)

Facial Recognition Vendor Test 2000 (FRVT 2000) Evaluation Report

Microsoft Research Vision Technology Group

The Face Recognition Home Page

Fingerprints:

FVC2004, Fingerprint Verification Competition

FVC2002, Fingerprint Verification Competition

An Introduction to Wavelets: FBI Fingerprint Compression

"Automated Systems for Fingerprint Authentication Using Pores and Ridge Structure," Proceedings of SPIE, Automatic Systems for the Identification and Inspection of Humans (SPIE Vol 2277), San Diego, 1994, p. 210-223. (802 kB PostScript)

FBI Fingerprint Compression Standard WSQ Software for UNIX Sun under SunOS 4.1.1 (uncertified)

Modern Communications Systems

FBI Fingerprint Compression Standard WSQ Software for Windows 3.1 (uncertified)

Free Software to Measure the Spatial Frequency Response (MTF) of Fingerprint Scanners

FVC2000, Fingerprint Verification Competition

NIST Visual Image Processing Group's Fingerprint Research

Signal Processing Research Center's Fingerprint Analysis

The FBI Fingerprint Image Compression Standard

Handwriting:

Document Understanding and Character Recognition (DIMUND)

Handwriting Recognition Group

Handschriften-E rkennungs-SYstem für Normalstifte (HESY) Signature Testing Device (works with normal pens)

Intelligent Recognition and Interactive Systems (IRIS), (The Nottingham Trent University)

The UNIPEN Project

Voice/Speech:

CAVE - The European Caller Verification Project

NIST Speaker Detection Evaluation

Modern Communications Systems

Speech Research, (Preception Science Laboratory, University of California)

Various:

Multimodal Verification for Teleservices and Security Applications (M2VTS)

Databases:

European Language Resources Association's (ELRA) Speech Databases

FVC2000, Fingerprint Verification Competition

Linguistic Data Consortium (voice)

M2VTS Multimodal Face and Voice Database

NIST Standard Reference Data

PEIPA, The Pilot European Image Processing Archive

YOHO (voice)

In testimony before Congress, the Biometrics Consortium told them that the methods they employ should be funded to help transfer of technology to the private and public sector systems that need the technology to ensure security.

Modern Communications Systems

21. GPS Locator Systems

   The Coast Guard runs 24 GPS satellites for sea navigation and the US

Department of Transportation (figures 65 & 66). They have a number of GPS

repeaters in the atmosphere to control the entire globe of GPS receivers on

ships, boats, land vehicles, commercial trucks, barges, passenger ships,

freighters, and US Navy ships. Many state highway department have GPS

ground stations on various towers in the states where they are located. GPS

stands for Global Positioning System and uses a digital signal to exactly position

the latitude and longitude of any point on earth. With this coordinate one can find

any position on the earth's surface. It is a standardized system and can be used

for hiking as well as other vehicle navigation. Handheld GPS locators are

available for $59 to under $300.00 on the open market. They are mighty

powerful when used in conjunction with a good topographical grid map in the

field. Many of the government surplus stores have good catalogues with GPS

devices in them too.

Figure 65. 24 GPS Satellite Network

Figure 66. NAVSTAR GPS Satellite by Rockwell International

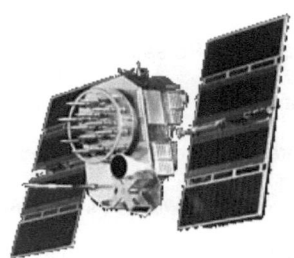

Modern Communications Systems

22. Cellular Systems

Cellular systems have changed the way we do business. Now drivers use cell phones on the highways, endangering others. Teenagers have cell phones to keep in touch with concerned parents. Businessmen use cell phones for routine calls when on the go. The number of cell systems has increased since 1988 by monumental amounts. Cell phones are linked to pagers. A missed call is a sin in the business world. The cell transmission covers more and more of America than ever before. The gaps between cells are becoming smaller and smaller so that wilderness areas are becoming covered as well as urban areas. The era of the personal device beckoned the development of the cell phone. It is the ultimate personal communications device rivaled only by the Star Trek TV series communicators. Marketing the convenience of cell phones seems to have done the trick. I believe they are good for emergency usage on the highway and teenagers should definitely have one when driving interstate or on large highways in vehicles which may not be so reliable. The #77 cell phone number for emergency use is a nice way for the public to stay in touch with the emergency services. However, the person who started the #77 number had no knowledge of cellular systems design. Cellular systems on the highway are backed up by the CB radios that truckers and travelers used in the 1970's during a popular marketing campaign by electronics companies. Citizen's Band radios still provide and excellent source of radio frequency communications in local vehicles and home base stations. The cellular system actually lets long distance be transmitted better than CB radios.

Cellular systems work in hexagonal cells in various urban and rural regions of the United States and operates in the 870 Mhz – 890 Mhz frequency range just above UHF TV channel frequency. If one travels, he passes through various cells that are interlinked. There are places in the United States that do not have cellular coverage or the signal becomes weaker. In fact there are places in the United States where FM radio does not transmit clearly (Oklahoma), unbelieveable as that sounds. Cellular towers relay the cellular signal broadcast to and from cellular telephones. There are also cellular radios. Cellular signals can be converted to transmission on other media too. A cellular phone call can be routed through a satellite channel for instance. Voice and data can be sent on cellular phones. An FBI agent or salesman in the field can transmit field reports via cellular phone system back to the field office. The data can then be forwarded to headquarters via land lines. This is called media transparency where the end user does not care how it gets to the recipient just that it gets there as fast as possible.

The cellular system will improve as the overall infrastructure improves in America to handle all telecommunications issues. Infrastructure issues include highways, schools, telecommunications, sewer and septic systems, email, satellite TV, radio waves, airlines routes in the sky, internet, urban developments

Modern Communications Systems

and sprawl control, cultural centers, public agency building locations, public real estate acquisitions, natural resources allocations and usage, environmental concerns, laws and regulations, local leadership, and funding through public and private means. The very same issues that America funded in the 1990's and are funding in Iraq are infrastructure issues involves communications systems.

23. Federal Telecommunications Regulations

The FCC is responsible for regulating communications including frequencies and management of communications. The best way to get to know the agency business is to read 47 CFR from the Government Printing Office in Washington DC. Communications systems acquisition is regulated by GSA in the FAR. OMB regulates computer and communications systems budgeting in OMB Circular A-130, A-109 and others. The DFAR is the Defense Federal Acquisition regulation and further muddies the waters by regulating the defense department acquisitions. We received many phone calls from confused contracting officers for the military who wondered if all the regulations apply to them that apply to other federal civilian agencies. The answer: A resounding affirmative! Even the companies have to conform to FCC - 47 CFR, GSA - 48 CFR, OMB, FAR – 48 CFR, DFAR and any other presidential memorandum on the subject if they do business with the federal government and each other in the private sector. For this reason, working in Washington for any amount of time in the policy making of technologies is a fascinating job and really opens your eyes to all the over-regulation that takes place in the industry. Congress keeps track of the infractions through the various Boards of Contract Appeals and the GAO who audits all agencies technologies funding. Many GAO reports talk about the mis-management of federal funds by the agencies, executive and independent. The word "taxpayer dollar" holds a lot of power in the Washington culture in reports and written documentation. GWU National Contract Law school is one of the specialty schools in DC that teaches all this information to willing students, most of whom are lawyers returning for refresher courses and new government contracting officers. The program is top notch and unique in the Nation. You better have a penchant for government history and logical thinking if you take these courses. I have met other contracting officers who have been educated there and we all agree it was the best in the government, bar none.

The FCC regulations writing process includes private sector comments on policy as well as internal comments processing similar to other regulatory agencies responsible for the CFR codifications of Congress. The GPO publishes the proposed regulatory changes for public comment. The public then has input to the regulations affecting them in 47 CFR. Appendix A includes some excerpts from 47 CFR.

Modern Communications Systems

24. Armed Forces Signal Processing Basics

The basic communications training for Armed Forces starts when one learns the letter alphabet for that service in basic training. Signal processing basics can be learned at one of the military signal schools or at local colleges. Museums are also a way to learn equipment capabilities and the history of the signals processing industry and government electronics in general. IEEE and AFCEA are two organizations that also give one a good footing in electronic systems and management. They both have professional development courses for experts in the field and members who want to learn more. IEEE recently added management courses to the plate for the palate. All the courses of study indicate the background of sciences at the high school level or above.

Figure 67. Courses of Study

Data Communications I & II

Networking

Cellular Phones Systems

Satellite Systems

Classified Systems

Project Management

Technical Systems Management

Modern Communications Systems

Figure 68. Types of Signal Processing Systems by Frequency

Cable TV

CB Radio

FM / AM Radio

Ham Radio

Aero-Communications

Harbor Navigation

VHF / UHF TV

Cellular Systems

Aircraft Radionavigation

Secure Telephone

Satellite

HDTV

Satellite Radio

Software Defined Radio

Museums –

One of the best ways to learn about systems are from the museums sponsored by the government and companies who display their products. Figure 69 shows some of the best museums for electronics equipment I have found in the Washington DC area. The equipment on displays usually tells the political story of how it was used by the government and who developed it and when they built it. There is usually no information on patents but this can be obtained from the patent office in Crystal City, Virginia or the University of Maryland Law School webpage in Baltimore.

Figure 69. Government and Military Museums with Electronics

National Electronics Museum

West Nursery Road

Linthicum, Maryland

National Cryptologic Museum

NSA

Seven Colony Road

Ft. Meade, Maryland

Test Pilots Museum

Patuxent River Naval Air Station

Lexington Park, Maryland

Smithsonian Air and Space Museum

Independence Avenue

Modern Communications Systems

Washington DC

AFCEA - The Armed Forces Communications and Electronics Association

Life membership costs $350.00. You receive Signal Magazine and can submit articles and learn more about Signals Intelligence and the military and contractor community. Some members of Congress and administrations past join the organization and give speeches to the members. They may also be on the board of directors who are many with connections through past positions. I have found the latest equipment needs are always advertised in the magazine for acquisition minded government people. They publish an annual book called Signals Source with phone numbers of all government contractors.

IEEE – Institute of Electrical and Electronics Engineers

This group is pretty good at research in electronics and computer systems. The publish many magazines in various areas of expertise. Spectrum Magazine is one you'll receive with membership. The are involved in standards writing for technical systems such as IEEE 802 for electronic networks. They also have training courses for professional development and one can volunteer as an officer if one likes. They are internationally recognized and sponsor some of the museums listed above. The Signals Processing group of IEEE covers many issues of concern to the military interception of signals intelligence. There is also an Intelligence group in IEEE that would be helpful in this area.

Modern Communications Systems

25. Transportation ITS Systems.

Intelligent Transportation Systems (ITS) are systems of systems that provide electronic capability to increase highway traffic efficiency. The mahjor urban areas in America and other G10 countries use ITS to augment traffic controls. An ITS may include video cameras, incident management, Commercial Vehicle Operations (CVO), Variable Message Signs (VMS), and Highway Advisory Radio (HAR). Telecommunications towers make this possible when they are placed along travel routes and major arteries. State Operation Centers control the highway incident response teams by using real time video camera feeds to see what is happening on the highways to cause delays and backups in congestion. Secondary crashes are reduced and personal property assets are saved as well as lives. Commercial Vehicle Operations use vehicle transponders to track trucks at weigh stations. Bypassing of weigh stations is possible fro enrolled trucks by weigh in motion detectors at the weigh stations. States are implimenting CVISN to track trucks at inspection stations. The advanced electronics and software makes this possible. CVISN is a system of systems that provide credentials, inspection, licensing, accident data, electronic screening, sensors, and wireless services.

Modern Communications Systems

26. HDTV and Satellite Radio

High Definition Television (HDTV) is a new technique that improves the quality of the picture on cable TV. The signal is run through the cable TV with a converter and wires to the HDTV ready LCD display. The picture quality and resolution is highly improved over normal TV. Sports games appear to be real time quality as if one weer sitting at the game with binoculars. The cost is more expensive than normal cable TV and normal cable signals can be delivered with HDTV signals on the same media (usually coax). Satellite HDTV is also available. Certain channels are provided as HDTV only channels. These are grouped together for easy reference during channel switching. Some channels are duplicated from normal cable and designated HD. The technology is very remarkable and provides viewing enjoyment.

Satellite Radio is broadcast signals from satellite receieved in a radio receiver. The channel bandwith is higher than normal radio and many more channels are available from all over the country in real time. The two main companies for satellite radio are XM Radio and Sirius. XM Radio is the older company and Sirius is up and coming. Each company has different sets of channels that they have negotiated to send signals to customers. The fee is about $10 a month to receive the signals through a special receiver antennae in the vehicle. Channels are grouped by topic. Weather channels can give weather

Modern Communications Systems

in any metropolitan area. Sports channels broadcast live events and recorded events and interviews. ESPN is broadcast from TV satellite and cable. The NFL has multiple games on the radio. Religious channels are available. And plenty of music is on satellite radio with various types of music played on different selected channels.

Both of these technologies has matured by the year 2006. They represent a new way of mass communications systems to the public in America. They are each more costly and higher quality than any preceeding technologies. It remains to be seen how each of these communications technologies will be further improved in the future which is very bright for both of them.

27. Project Management Considerations

There are many project managers of modern communications systems who have normal project management skills mixed with telecommunications engineering knowledge. A good communications project manager understand how to manage frequencies. He also understands the details of each phase of implementation of the communications system he is installing using either classical project management watershed approach or some of the newer project management principles such as Spiral model or Rapid Prototyping. The watershed model seems to work best in most situations across technical building projects. A good approach should include using some software to manage the project life cycle such as MS Project, Mac Project or other software. I have seen telecommunications Project Managers whose skills were very portable and they moved from one project in government to projects in the private sector rather easily. Figure 70 shows the typical project management life cycle phases in the Watershed model as applied to a major telecommunications project.

Figure 70. Watershed model for Telecommunications System

1. Feasibility Phase or Stage |-----|
2. Requirements Analysis / Definition |-----|
3. System Design |-----|

Modern Communications Systems

4. System Installation and Programming |-------|

5. System Testing |-----|

6. System Operations & Maintenance |-------|

The concern is for the project resources and making sure resources are available from an IRM perspective. The funding precludes the people and contractual staff and major physical assets of the project. Resources are usually scarce on most projects and the project manager has to ensure funding is available throughout the life cycle of the project as his number one priority.

Some examples of some project managers whom have done very good jobs on government communications contracts are FTS 2000, C-SPAN, DNS, and Internet 2. The government work can not be done without teamwork from the contractors on all tasks in the watershed model. Government staff are trained at managing the installation and many can do some of the tasks or have done them in the private sector before they manage the project. It may be the case that the PM has learned his skills from college or technical school or a parent. The best PM's are very flexible and not rigid in their approach to completing tasks unless there is a time schedule that must be adhered to. A tight time schedule might mean that the PM has to crash the network critical path to complete the project under the CMP/PERT methodology. CMP / PERT was first used on the Polaris Missile System in the late 1950's and stands for Critical Path Method / Performance Review Evaluation Technique. It has been taught in the DOD

Modern Communications Systems

systems and project management classes for years. Every major military organization uses CPM / PERT and decision point papers to document systems development along with DOD directives, Navy, Air Force, or Army directives. The Marines Corps use Navy Directives for all systems building and management. The Federal, state, and local governments uses FIPS Publications and NIST publications as guidance and communications industry standards such as IEEE standards. The Federal Communications Regulations are good for understanding the Federal Regulations (47 CFR) of the telecommunications industry. Industry and agencies are allowed to comment on the regulations as they are developed by FCC.

Alternative methods of project management are rapid prototyping (Figure 71), spiral method, and automated methods including software such as MS Project, MAC Project, and Superproject. Lois Zells in her book Managing Software Projects describes systems that help with project management. They should be used practically as a support to the actual software builds by the project leader to define where the project is going and how the status reporting on the projects is completed. Status reports done weekly are important to hold people on the project accountable for results in building communications systems. Phases withing these software sets are handled very similar to the watershed model. The resources used in the software are sometimes "leveled" and balanced for the number of hours the persons work on the projects.

Figure 71. Rapid Prototyping

Phase I - Requirements Analysis

Phase II – Working Model Presented to End User

Phase III – Interative Refinements Made to Working Model

Phase IV – Final System Accepted

Rapid prototyping works expecially well for the software portions of building communications projects. Watershed project management works well for constructing the hardware platform and wiring of the communications system because it came from the construction industry where hard products are visible and can easily be measured. Software is harder to see and feel and thus is better produced using rapid prototyping in situations where you have to get the requirements correct the first time from the end user on software systems that are built to be highly interactive and user friendly such as an email software package. Off the shelf software is now used most frequently to meet needs where in house developer do not have the unique requirements to do generic computerized communications, database, spreadsheet, drawing, CADD, charting, graphing, presentations, pictures and other job functions and tasks that are available in the marketplace at reduced costs savings. Sometimes for those who are no technically savvy it is recommended to also buy technical support with the off the shelf software at least as a phone call for problem resolution.

28. Acquisition and Planning Considerations

The very best government acquisition training is available from various military organizations and university programs. Many of these schools are in Washington DC or around the country. Figure 72 shows the most famous ones. The George Washington National Law Center Contract Law program is very good. One learns how to do all the contract management functions in this program. Communications systems are acquired through government management of funds at GSA as part of the government infrastructure. State government handles technology for communications systems on an agency by agency basis but has no technology management agency per se. Major corporation acquire their own private networks or use commercially available systems that are already in place.

Figure 72. Government Systems Acquisition Schools

1. National Defense University

2. Fort Belvoir Army Acquisition School

3. George Washingtion University

4. Naval Postgraduate School

5. Air Force Institute of Technology

6. Other Graduate School MBA and IRM Programs

7. GSA 1000 by 2000 Program – Trail Boss

Figure 73. GWU Courses in Acquisition Training

1. Introduction to Contracting

2. Contract Administration

3. Government Contract Law

4. Government Contract Audits

5. Operating Practices in Contract Administration

6. Contract Pricing

7. Contract Negotiation

8. Cost Reimbursable Contracts

9. Intellectual Propoperty, Patents, and Software

10. Source Selection: Best Value

11. Contract Economics

12. Contract Disputes and Terminations

13. Construction Contracting

14. FAR Part 15

15. Federal Appropriations Law

16. Patent, Technical Data, and Computer Software

17. Procurement and the Internet

18. Project Management for Contracting Professionals

19. Subcontract Management

20. Task Ordercontracting

21. Understanding the Cost Accounting Standards

22. Simplified Acquisitions

23. International Contracting

24. Performance based Service Contracts

25. Cost Estimating

26. Business Law and the UCC

27. Advanced Procurement Issues

28. Advanced Contract Administration

Source: ESI International, 2004

Figure 74 shows the Federal boards of contract appeals that companies can refer to for contract cases. Some larger companies dispute an average of 10% of their lost contracts in hopes of winning an appeal based on a mistake by the government contracting officer. Some companies do not deal with government contracts since they do not have as high a return as private sector commercial contracts. A case can go to appeal in several board of appeals if there is no remedy.

Modern Communications Systems

Figure 74. Federal Contract Boards of Appeals

1. GSA Board of Contract Appeals

2. GAO Board of Contract Appeals

3. Army Board of Contract Appeals

Modern communciations systems acquisition requires the program manager to understand the technical and non technical issues on communications systems implementation projects as well as project management. Effective program management is defined as managing many projects simulataneously. Major military and federal communications systems project acquisition comes up for review in Congress during the appropriations process. The process is an annual one in which Congress provide money to do the business of the government. Nagle's book[18] discusses the process as it has taken place since the Revolutionary War in America. Nagle was on staff at George Washington University when he wrote this book. Professors Nash and Cibinic run the George Washington program and are very good at it. Many Congressmen send their staff to the classes at George Washington to better understand the processes in government contracting and the law as it exists and it's history. It helps them make decisions on future appropriations. GW has graduated over 1800 people from this program. The federal government has more than 22,000 contracting officers and many more people who serve as COTR on government projects.

---

[18] Nagle, History of Government Contracting

Modern Communications Systems

The bible for contracting officers is the FARS and DFARS and agency directives. OMB circulars also help give directions for managers from the executive branch. A person involved in acquisition needs to know what the NIST and IEEE standards say for the acquisition involving communications systems. One does not need to know all the details to obligate the funds but more information definitely helps. Writing the regulations for acquisition is a matter of fully understanding every aspect of telecommunications in government services and the private sector and experience directing projects and people.

A contract well written is worth it's weight in gold. The contract should be equitable and not cheat the contractor people out of time and money on certain contracts or staff will leave the contract. Agency procurement rules should be reviewed before writing a contract bid document in an agency. State government has a set of it's own procurement rules and thresholds for dollar values that must be followed in a procurement. The larger the procurement the more bidding that needs to be done in order to save money. Three bids are good on small procurements. An open bid process is held for larger projects. Construction projects have an Architectural and Engineering contract bid process. The main concern during the paperwork and meetings should be keeping integrity of the process. A good WBS, SOW, and other contract details make the project go smoother. The biggest problems is that CO's don't know about everything they are acquiring and some need to learn at the time of writing the contract which canb cause technical errors later in the project. Many CO's have been project managers previously. Another aspect of contract officers tasks is market research which gives them information to use on specifying contract deliverables. The deliverable must be fully described in detail in one type of specification. It's function can be described or it's output's and design can be specified. For telcom systems this means getting the best telcom system design for the government leaving some decisions up to the contractor during implantation. Hardware needs to be described exactly as possible. Most contractors who are in the business can tell one what you need in the contract for that technology and you are allowed to ask them as long as you do not write the contract just for them and openly bid it in a sealed bid process.

29. Seven Classical Management Principles Applied to Today's Technology

Good management is a matter of practice as well as academic studies. Once a manager has the keys to understanding what good management principles are considered useful for the government and private sector, he has an obligation to teach them and use them wisely with people. My comprehensive exams included an essay on this theme of applying the principles of classical management to the project management of an agency computer system in Washington DC. The test question was seeking if I knew how to understand and apply basic management knowledge and I am happy that I passed with flying colors by adapting from the Ed Yourdon book "Life Cycle Management Principles" and these seven measures of classical management. Together these provide a framework for technical project management in the 21st century. Interestingly enough the military calls the control and communications functions of management C4I - Command, control, communications, computers and Intelligence. This has evolved over the years from C2 or command and control.

The seven classical management principles according to the Sisk book and Peter Drucker, the Claremont Graduate School management professor are:

Modern Communications Systems

Figure 75. The 7 Classical Management Principles

1. Planning

2. Organizing

3. Control

4. Communications

5. Directing

6. Evaluating and

7. Staffing.

These seven functions are the gist of what good manager's do with their time on all projects. I memorized these functions as POCCDES acronym and reproduced them for my comprehensive exam for my masters degree at American University. I explained how they related to the functions of the normal project manager's job on technology project. Only later did I start using these functions on my own projects with an eye towards guiding my own projects and staff toward successful completion. The seven are fool proof. The are the meat and potatoes of the basic principles of management as I took in an undergraduate course at St. Mary's College of Maryland back in 1976. In that class I decided I wanted to be a systems technology manager someday for two reasons 1 - I was also taking computer science courses in math and 2 – my father encouraged me to be better than his manager's at the place he worked as an electrical engineer. In fact, these seven functions have been time tested by many of the best managers in most businesses. The text used in that Principles of Management class was the Sisk text[19] used by many other colleges. I liked the fact that he described IT (then called ADP) as making the organization more competitive as information is the lifeblood of both the communication function and the business intelligence functions. Learning what the competition is doing is key to any businesses survival as well as government unit survival. Today, there are many new management methods and styles which are also described in Sisk, but these seven are the basics that one needs to learn before all the other styles and management rational techniques that one learns in analytical and numerical courses on economics and project management methods. I believe that if these seven functions were good enough to win me my masters degree than they must be important to all managers jobs whether they are task masters or people oriented managers. The management grid's are also popular for this dichotomy. Factor in Abraham Maslow's hierarchy of employee needs and the Meyers Briggs Type indicator and now you've got a manager ready to face the real world with some useful tools in his chest. These have been the most useful tools I could ever have had in any formal education I had received for my government career as a manager. They apply directly to the manager developing himself and understanding what motivates other people which varies from person to person. There are a lot or project management skills that come in handy but many of these are automated in software today and can be learned in class or in the software daily usage. The rational style of education will yield managers who can make a profit for the company and maximize Returns on

[19] Sisk, Management and Organization, 2nd edition, South Western Publishing, 1973.

Modern Communications Systems

Investments and Internal Rates of Return. These are very useful for both the new government of the 2000's as well as the private sector. Managers from the 1960's and 1970's models can not relate to these concepts because they are trapped in the models of cause and effect of those times. After Reagan became president the models evolved for government service and government became a more privatized organization with greater ties to the private sector for normal business. Normal government functions became cost based to government employees rather than free benefits. A new federal and state retirement system moved employees to more of a system based on investments than one based on a percentage of contributions. This had the effect of cheating the younger employees when compared to older retirement plans.

As a practical application, if we take the seven classical management components and map them into our daily lives, we find that our actions will bear out the fact that we are constantly applying one of the seven classical management principles most of the time in our work. MIS and this book is concerned with primarily the communications component of the seven classical management functions. The Sisk book tells us that information is required when in a competitive business situation. Well the communications function helps people be more competitive and do their jobs better and more efficiently just as the applications programming function helps managers understand their tactical and operational data better. Communications with a goal in mind and purpose is always better than that with none in mind. When the goal is met, the communications has been productive in the business sense. When it is not met, we have to increase our desire to succeed and commucniate better like in person.

However, there can be abuses to the communications functions. When an email software user starts using emails to dominate the environment he has gone to an extreme of electronic communications and tried to overcontrol his environment with too much communications. If the feedback (from all sources) from the excess communications is poor or bad then an adjustment is required. Today, there is nothing I hate worse than an email left un-replied. But people have different schedules and this is to be expected. The percentage of unreplied emails is a measure of communications effectiveness and feedback in itself.

Telecomm Systems Economics –

The ecomonics logic of communications systems goes like this. For a LAN, If we can share computer resources on a network and use them more efficiently then we save money. By installing a smaller network of highly capable machines in replace of the old mainframe architectures we downsize and improve the processing power of the average person. We call this server technologies and the return is higher when people share printers, plotters, and other resources. Long haul telecom is best competed by many companies and not monopolized by any one company. The 1984 decision deregulated Telecom

and provided for competitive pricing by companies, that is if you believe they will

try to undercut each other to obtain more business from customers. Some

government telcom systems are private type networks as used by the military

and the bidders compete but the winner may also receive a follow maintenance

contract for good performance on the contract. Economies of scale in telcom

systems means processes as many messages as a channel can handle and

having a very high capacity for more customers on the network. This is

monitored by the network center and is known as network traffic management.

The system as a high return on investment if it operates at close to full capacity

and is very reliable.

Reliability and Availability -

Reliability calculations and availability (uptime) can be computed for the

system based on previous experience. Reliability is increased by redundancy in

systems circuits like parallel circuits rather than serial circuits. Obviously, we

shoot for 100% uptime but realistic uptime may be more like 98% or 99.5% of the

time. More highly reliable systems cost more money to install because they are

installed in parallel and have redundancy built into the system. Messages are

allowed to roll over to the new system on the fly to take advantage of the survival

of the network switches. Internet works like this and there is survivability built into the routing of the messages which are broken down and sent by any possible means. The cost for backup switches is minor because there are multiple routes a message can take to the destination. What we are really talking about is maintaining high quality of service.

High Quality Service –

Maintaining quality in communications systems is a matter of increasing reliability and availability while improving overall quality of service to the end users of the communications systems. How often has a communications company done a survey of quality of service for you in your lifetime? Again how often have they offered a discount of their services to beat the competition for a short period and then jacked the price back up higher than before? I have found that once the various ISP's get used to your business, they all ignore you and ignore any rules they have to obtain new business. They essentially forget their old reliable customers and often charge increases in service. When was the last decrease in service you got passed along to you for any communications service? I know high quality when I see it in computer systems and communications services. I once had a company try to double charge me for services for 6 months until I saw the bills and demanded a refund. They acted like I was wronging them and did not send me any refund. Rather they thought they were smarter by making me take free service for 6 months. I changed ISP service company at the end of the time period on principle of the matter. Do you think they would tell other customers how they overcharged me? I should have told Congress or the BBB if I thought they were effective enough at fixing this behavior. You need to demand high reliability and high availability to run a high quality system of any sort. Low price doesn't hurt either, but be careful that if you just go by cost, you will not get the highest level of qulity for the application you are running. I learned this when I switched over to cable modem service from 56K baud service. You just can not make a slow service act like a faster service unless you pay the price for better quality.

Modern Communications Systems

30. The Future of Communications

This chapter discusses the future of communications and possible changes we may see in the next years. Figure 76 shows the historical timeline for some major human communications since the beginning of time.

Figure 76. Communications Timeline[20]

BC Caveman drawings on wall and primitive languages

BC Babylonian Tablets

BC Tower of Babel (many languages in mankind)

BC Egyptian Papyrus & Cryptography

BC Torah – Pentatuch (5 books of Moses)

200 AD Roman Appian Way messages

350 AD 66 Chapter Bible Written

800 AD Koran

900 AD Discovery of Dead Sea Scrolls

1100 AD Gnostic Gospels Discovered in Hammerabai

1400 Gutenberg Modern Printing Press Invented

1837 Samuel Morse develops Morse Code and Telegraph

1867 First Typewriter in US

1876 First Telephone – Alexander Graham Bell

1893 First Motion Picture – Thomas Edison

---

[20] Time Almanac 2003, Time Home Entertainment, 2003

Modern Communications Systems

1901 First Transatlantic Radio developed by Marconi

1934 AT&T Wired America's Telephones

1936 First Mass Communication Radio Speech by Hitler

1945 First Mainframe Computer ENIAC – Ekert and Mauchley

1951 First US Tv Broadcast

1959 First Communications Satellite

1969 DARPA Internet Project

1972 First Video Disk - Netherlands

1976 President Jimmy Carter sends first email

1977 First Personal Computer

1983 First Cable TV Station

1988 Cellular Phones

1993 Commercial Internet ISPs

1999 Satellite Radio

The Personalization of Communications (smaller systems)

In the last 10 years manufacturers have personalized communications by making all types of devices that are personal for items that could be shared previously. The move towards more mobility and smaller hardware has infused new sales in these devices. The microchip and VLSI miniaturization has helped make this possible and embedded computers have also helped make this possible. Also the internet infrastructure has now become universal and people are using more cellular systems than ever before according to the World

Modern Communications Systems

Almanac 2002. Government systems that are tested and work are then rolled out for the rest of America to use full time in commercial mode.

.

New Marketing Strategies (more integrated, in cars, everywhere we go, new technologies, kids)

New marketing of technologies has increased the number of technologies we use in everyday life. Companies are making profits off the new technologies by not obsoleting the previous technologies. Communications technologies that are very highly effective tend to stay in the public for a while. Electronic communications are usually the best way we communicate in today's society. New communications technologies marketing strategies are focused on the places we travel in cars, boats, and other vehicles. They are also focused on our children. New communications technologies focus on the family on TV commercials and try to make us believe we need them.

Where man will go in space (Mars colonies)

In the future new communications technologies willl go where man goes in outer space. NASA will send deep space probes to the far planets and when men go there, we will colonize the far reaches of the galaxy.

New discoveries in sciences related to communications

We will discover new sciences related to communications as we venture out into space on other planets. New research and development will be done

Modern Communications Systems

into areas that will positively affect the sciences of communications. Some discoveries that could be made soon:

1. A new material for propagating electronic media from other planets rocks
2. New Frequency Spectrum on thehigh end of the chart
3. More new computer software advances
4. More Computer Hardware advances (which drive software advances)
5. New methods of transportation (communications has always codciled with transportation improvements and supplanted them)

New human physiology computer interactions

New human physiology and computer interactions will cause humans to better understand how we can augment our capabilities in communications. Computers can help us build a better world for communications. Biomedicine can contribute greatly in the following areas:

1. Discovering and unlocking the hidden potentials of the human brain such as telepathic communications
2. Discovering new genes that make us smarter to produce more new devices and inventions.
3. Pushing the human body to its limitations
4. Improving our human – machine brain power and interaction (nueral networks)
5. Increasing our life spans

Modern Communications Systems

New electronics and engineering improvements

IBM has the most research patents yearly of any company in high technology. New technology is available every year with new young minds producing new ideas that improve our telecommunications industry. It is reasonable to think that we will invent many new devices that will be marketed to the public in the future. Miniaturization of components and VLSI led to the smaller systems that are marketed as personal devices today. Economics says that higher income families will use these technoloigies as conveniences if they can afford them. Electronics companies know this and target these upper income families as well as children. Military is targeted by ex-DOD people who now work for private sector defense contractors. Military advancement often lead to other advancements in the public technologies. One History Channel show stated that Man's wars often lead to advancements in sciences that are peaceful such as atomic energy plants coming out of the atom bomb research. The show "Tactical to Practical" is all about this transition of technology. Even Da Vinci created wartime devices when he thought of one such as the catapult and helicopter. He used his genius to influence many aspects of modern society and protect his country. The Germans in World War II and the years preceeding this war did the same thing. America is always developing new technologies because we are a free country who encourages new ideas. We have the ability to field many new telecommunications systems in the future from private and academic sources. Our research schools are the best in the world. Maintaining the highest level of competition requires us to bring the best minds from the world

Modern Communications Systems

to America to help our causes. It has been this way in the 20th century and will

be even more critical in the future to fight terrorism and other evils that afflict

America. As the honorable Tom Ridge stated, this new generation has the

challenge of fighting terrorism, just like the previous generation fought the cold

war, and the greatest generation fought World War II. . No one likes wars who is

a rational thinker. Thomas Malthus believed that war kept the population down

with pestilence and disease in mankind (in the old world). There is no reason we

should believe we have to always fight in wars, yet this seems to be man's

destiny. The Tower of Bable and God's creation of multiple languages divided

mankind forever. If we had more insights and understanding of various

languages and cultures would we fight each other less? Maybe. Did God mean

for us to communicate with each other and still retain our own identities? The

Pope knows many different languages and makes an honest attempt to

understand others in their languages. It's a sad thing that most people are

tempted to hate others because they are unlike them rather than communicating

with them in honest dialogue. This is the basis of diplomacy and foreign

relations.

Modern Communications Systems

Appendix A.

Telecom Standards and Regulations

The following list details a sampling of major telecommunications standards and regulations in the United States and Europe.

U.S. Telecom Regulations

FCC Licensed Radio Standards—CFR 47

Part 5 Experimental radio service (other than broadcast).

Part 6 Access to telecommunications service, telecommunications equipment, and customer premises equipment by persons with disabilities.

Part 7 Access to voice mail and interactive menu services and equipment by people with disabilities.

Part 11 Emergency Alert System (EAS).

Part 15 Radio frequency devices.

Part 18 Industrial, scientific, and medical equipment.

Part 20 Commercial mobile radio services.

Part 21 Domestic Public Fixed Radio Services; Subpart K: Multipoint Distribution Service (MDS).

Part 22 Public Mobile Services; Subpart E: Paging and Radiotelephone Service; Subpart F: Rural Radiotelephone Service; Subpart G: Air-Ground Radiotelephone Service; Subpart H: Cellular Radiotelephone Service; Subpart I: Offshore Radiotelephone Service.

Part 23 International Fixed Public Radiotelephone Services.

Part 24 Personal Communications Services; Subpart D: Narrowband PCD; Subpart E: Broadband PCS.

Part 25 Satellite Communications.

Part 26 General Wireless Communications Service.

Part 27 Wireless Communications Service.

Modern Communications Systems

Part 73 Radio Broadcast Services; Subpart A: AM Broadcast Stations; Subpart B: FM Broadcast Stations; Subpart C: Noncommercial Educational FM Broadcast Stations; Subpart E: Television Broadcast Stations; Subpart F: International Broadcast Stations; Subpart G: Emergency Broadcast System.

Part 74 Experimental Radio, Auxiliary Special Broadcast, Other Program Distributional Services; Subpart A: Experimental Broadcast Stations; Subpart D: Remote Pickup Broadcast Stations; Subpart E: Aural Broadcast Auxiliary Stations; Subpart F: TV Broadcast Auxiliary Stations; Subpart G: Low-Power TV, TV Translator, and TV Booster Stations; Subpart H: Low-Power Auxiliary Stations; Subpart I: Instructional Fixed Service; Subpart L: FM Broadcast Translator and Booster Station.

Part 76 Multichannel video and cable television service.

Part 78 Cable Television Relay Service.

Part 80 Stations in the Maritime Services; Subpart J: Public Coast Stations; Subpart K: Private Coast Stations and Marine Utility Stations; Subpart L: Operational Fixed Stations; Subpart M: Stations in the Radiodetermination Service; Subpart N: Maritime Support Stations; Subpart O: Alaska Fixed Stations; Subpart V: Emergency Position Indicating Radiobeacons (EPIRBs); Subpart W: Global Maritime Distress and Safety System (GMDSS).

Part 87 Aviation Services; Subpart F: Aircraft Stations; Subpart G: Aeronautical Advisory Stations (Unicoms); Subpart H: Aeronautical Multicom Stations; Subpart I: Aeronautical Enroute and Fixed Stations; Subpart J: Flight Test Stations; Subpart K: Aviation Support Stations; Subpart L: Aeronautical Utility Mobile Stations; Subpart M: Aeronautical Search and Rescue Stations; Subpart N: Emergency Communications; Subpart O: Airport Control Tower Stations; Subpart P: Operational Fixed Stations; Subpart Q: Stations in the Radiodetermination Service; Subpart R: Civil Air Patrol Stations; Subpart S: Automatic Weather Observation Stations.

Part 90 Private Land Mobile Radio Services; Subpart B: Public Safety Radio Pool; Subpart C: Industrial/Business Radio Pool; Subpart F: Radiolocation Service; Subpart J: Nonvoice and Other Specialized Operations; Subpart M: Intelligent Transportation Systems Radio Service; Subpart P: Paging Operations.

Part 95 Personal Radio Services; Subpart A: General Mobile Radio Service (GMRS); Subpart B: Family Radio Service (FRS); Subpart C: Radio Control (R/C) Radio Service; Subpart D: Citizens Band (CB) Radio Service; Subpart F: Interactive Video and Data Service (IVDS); Subpart G: Low-Power Radio Service (LPRS).

Modern Communications Systems

Part 97 Amateur Radio Service.

Part 100 Direct broadcast satellite service.

Part 101 Fixed Microwave Services; Subpart G: Digital Electronic Message Service; Subpart H: Private Operational Fixed Point-to-Point Microwave Service; Subpart I: Common Carrier Fixed Point-to-Point Microwave Service; Subpart J: Local TV Transmission Service; Subpart L: Local Multipoint Distribution Service (LMDS).

Note: Only some of the subparts contained within each part are listed above, to illustrate the distinctly different services regulated under each part. Each part also contains other subparts, including the general technical standards applying to all subparts.

European Telecom Standards

R&TTE Directive Standards

Harmonized standards for the implementation of Directive 1999/5/EC of the European Parliament and of the Council of 9 March 1999 on radio equipment and telecommunications terminal equipment and the mutual recognition of their conformity. April 2000. 2000/C 99/02.

ETSI

EN 301 419-1 V4.1.1 (04-2000)—Digital cellular telecommunications system (Phase 2); Attachment requirements for global system for mobile communications (GSM); Part 1: Mobile stations in the GSM 900 and DCS 1 800 bands; Access (GSM 13.01 version 4.0.1).

EN 301 419-2 V5.1.1 (04-2000)—Digital cellular telecommunications system (Phase 2+); Attachment requirements for global system for mobile communications (GSM); High-speed circuit switched data (HSCSD) multislot mobile stations; Access (GSM 13.34 version 5.0.3).

EN 301 419-3 V5.0.2 (11-1999)—Digital cellular telecommunications system (Phase 2+); Attachment requirements for global system for mobile communications (GSM); Advanced speech call items (ASCI); Mobile stations; Access (GSM 13.68 version 5.0.2 Release 1996).

EN 301 419-7 V5.1.1 (09-2000)—Digital cellular telecommunications system (Phase 2+); Attachment requirements for global system for mobile communications (GSM); Railways band (R-GSM); Mobile stations; Access (GSM 13.67 version 5.0.2).

Modern Communications Systems

EN 301 426 V1.2.1 (10-2001)—Satellite earth stations and systems (SES); Harmonized EN for low data rate land mobile satellite earth stations (LMES) operating in the 1.5/1.6 GHz frequency bands covering essential requirements under Article 3(2) of the R&TTE Directive.

EN 301 427 V1.1.1 (05-2000)—Satellite earth stations and systems (SES); Harmonized EN for low data rate land mobile satellite earth stations (LMES) operating in the 11/12/14 GHz frequency bands covering essential requirements under Article 3(2) of the R&TTE Directive.

EN 301 428 V1.2.1 (02-2001)—Satellite earth stations and systems (SES); Harmonized EN for very small aperture terminal (VSAT); Transmit-only, transmit/receive or receive-only satellite earth stations operating in the 11/12/14 GHz frequency bands covering essential requirements under Article 3(2) of the R&TTE Directive.

EN 301 430 V1.1.1 (05-2000)—Satellite earth stations and systems (SES); Harmonized EN for satellite news gathering transportable earth stations (SNG TES) operating in the 11–12/13–14 GHz frequency bands covering essential requirements under Article 3(2) of the R&TTE Directive.

EN 301 441 V1.1.1 (05-2000)—Satellite earth stations and systems (SES); Harmonized EN for mobile earth stations (MES), including handheld earth stations, for satellite personal communications networks (S-PCN) in the 1.6/2.4 GHz bands under the mobile satellite service (MSS) covering essential requirements under Article 3(2) of the R&TTE Directive.

EN 301 442 V1.1.1 (05-2000)—Satellite earth stations and systems (SES); Harmonized EN for mobile earth stations (MES), including handheld earth stations, for satellite personal communications networks (S-PCN) in the 2.0 GHz bands under the mobile satellite service (MSS) covering essential requirements under Article 3(2) of the R&TTE Directive.

EN 301 443 V1.2.1 (02-2001)—Satellite earth stations and systems (SES); Harmonized EN for very small aperture terminal (VSAT); Transmit-only, transmit-and-receive, receive-only satellite earth stations operating in the 4 GHz and 6 GHz frequency bands covering essential requirements under Article 3(2) of the R&TTE Directive.

EN 301 444 V1.1.1 (05-2000)—Satellite earth stations and systems (SES); Harmonized EN for land mobile earth stations (LMES) operating in the 1.5 GHz and 1.6 GHz bands providing voice and/or data communications covering essential requirements under Article 3(2) of the R&TTE Directive.

EN 301 459 V1.2.1 (10-2000)—Satellite earth stations and systems (SES); Harmonized EN for satellite interactive terminals (SIT) and satellite user terminals (SUT) transmitting towards satellites in geostationary orbit in the 29.5 to 30.0 GHz frequency bands covering essential requirements under Article 3(2) of the R&TTE Directive.

EN 301 502 V8.1.2 (07-2001)—Harmonized EN for global system for mobile communications (GSM); Base station and repeater equipment covering essential requirements under Article 3(2) of the R&TTE Directive (GSM 13.21 version 8.0.1 Release 1999).

EN 301 511 V7.0.1 (12-2000)—Global system for mobile communications (GSM); Harmonized standard for mobile stations in the GSM 900 and DCS 1800 bands covering essential requirements under Article 3(2) of the R&TTE Directive (1999/5/EC) (GSM 13.11 version 7.0.0 Release 1998).

EN 301 681 V1.3.2 (01-2003)—Satellite earth stations and systems (SES); Harmonized EN for mobile earth stations (MES) of geostationary mobile satellite systems, including handheld earth stations, for satellite personal communications networks (S-PCN) in the 1.5/1.6 GHz bands under the mobile satellite service (MSS) covering essential requirements under Article 3(2) of the R&TTE Directive.

EN 301 721 V1.2.1 (06-2001)—Satellite earth stations and systems (SES); Harmonized EN for mobile earth stations (MES) providing low bit rate data communications (LBRDC) using low earth orbiting (LEO) satellites operating below 1 GHz covering essential requirements under Article 3(2) of the R&TTE Directive.

EN 303 035-1 V1.2.1 (12-2001)—Harmonized EN for TETRA equipment covering essential requirements under Article 3(2) of the R&TTE Directive; Part 1: Voice plus data (V+D).

EN 303 035-2 V1.2.2 (06-2003)—Harmonized EN for TETRA equipment covering essential requirements under Article 3(2) of the R&TTE Directive; Part 2: Direct mode operation (DMO).

Note: In addition, standards published under Directives 73/23/EC and 89/336/EEC may be used to demonstrate compliance with Articles 3(1)(a) and 3(1)(b) of Directive 1999/5/EC.

Digital Enhanced Cordless Telecommunications (DECT) Standards

ETSI

EN 300 765 V1.3.1 (04-2001)—Digital enhanced cordless telecommunications (DECT); Radio in the local loop (RLL) access profile (RAP); Part 1: Basic telephony services.

Modern Communications Systems

EN 300 765-2 V1.2.1 (02-2001)—Digital enhanced cordless telecommunications (DECT); Radio in the local loop (RLL) access profile (RAP); Part 2: Advanced telephony services.

EN 301 240 V1.1.3 (06-1998) (Historical)—Digital enhanced cordless telecommunications (DECT); Data services profile (DSP); Point-to-point protocol (PPP) interworking for Internet access and general multiprotocol datagram transport.

EN 301 242 V1.2.2 (09-1999)—Digital enhanced cordless telecommunications (DECT); Global system for mobile communications (GSM); DECT-GSM integration based on dual-mode terminals.

EN 301 406 V1.5.0 (02-2003)—Digital enhanced cordless telecommunications (DECT); Harmonized EN for digital enhanced cordless telecommunications (DECT) covering essential requirements under Article 3(2) of the R&TTE Directive.

EN 301 649 V1.3.1 (03-2003)—Digital enhanced cordless telecommunications (DECT); DECT packet radio service (DPRS).

EN 301 650 V1.2.1 (04-2002)—Digital enhanced cordless telecommunications (DECT); DECT multimedia access profile (DMAP); Application-specific access profile (ASAP).

Other Telecom Standards

ATIS T1.107:2002—Telecommunications—Digital hierarchy—Formats specifications.

ATIS T1.413:1998—Asymmetric digital subscriber line (ADSL) metallic interface.

ATIS T1.413a:2000—Telecommunications—Network and customer installation interfaces—Asymmetric digital subscriber line (ADSL) metallic interface (supplement to ATIS T1.413:1998).

EN 300 652:1998—Broadband radio access networks (BRAN); High-performance radio local-area network (HIPERLAN) Type 1; Functional specification.

EN 60244-5:1994—Methods of measurement for radio transmitters—Part 5: Performance characteristics for television transmitters.

EN 60244-8:1994—Methods of measurement for radio transmitters—Part 8: Performance characteristics of vestigial-sideband demodulators used for testing television transmitters and transposers.

EN 60244-9:1994—Methods of measurement for radio transmitters—Part 9: Performance characteristics for television transposers.

EN 60244-10:1993—Methods of measurement for radio transmitters—Part 10: Methods of measurement for television transmitters and transposers employing insertion test signals.

EN 60244-11:1993—Methods of measurement for radio transmitters—Part 11: Transposers for FM sound broadcasting.

EN 60244-13:1993—Methods of measurement for radio transmitters—Part 13: Performance characteristics for FM sound broadcasting.

EN 60244-14:1997—Methods of measurement for radio transmitters—Part 14: External intermodulation products caused by two or more transmitters using the same or adjacent antennas.

EN 60244-15:2000—Methods of measurement for radio transmitters—Part 15: Amplitude-modulated transmitters for sound broadcasting.

EN 60315-3:2000—Methods of measurement on radio receivers for various classes of emission—Part 3: Receivers for amplitude-modulated sound-broadcasting emissions; Amendment A1:1999 to EN 60315-3:1999.

EN 60835-1-3:1995—Methods of measurement for equipment used in digital microwave radio transmission systems—Part 1: Measurements common to terrestrial radio-relay systems and satellite earth stations—Section 3: Transmission characteristics; Amendment A1:1995 to EN 60835-1-3:1995.

EN 60835-1-4:1995—Methods of measurement for equipment used in digital microwave radio transmission systems—Part 1: Measurements common to terrestrial radio-relay systems and satellite earth stations—Section 4: Transmission performance; Amendment A1:1995 to EN 60835-1-4:1995.

EN 60835-2-1:1994—Methods of measurement for equipment used in digital microwave radio transmission systems—Part 2: Measurements on terrestrial radio-relay systems—Section 1: General.

EN 60835-2-2:1995—Methods of measurement for equipment used in digital microwave radio transmission systems—Part 2: Measurements on radio-relay systems—Section 2: Antenna.

EN 60835-2-3:1995—Methods of measurement for equipment used in digital microwave radio transmission systems—Part 2: Measurements on terrestrial radio-relay systems—Section 3: RF branching networks.

EN 60835-2-7:1995—Methods of measurement for equipment used in digital microwave radio transmission systems—Part 2: Measurements on terrestrial radio-relay systems—Section 7: Diversity switching and combining equipment.

EN 60835-2-8:1996—Methods of measurement for equipment used in digital microwave radio transmission systems—Part 2: Measurements on terrestrial radio-relay systems—Section 8: Adaptive equalizer; Amendment A1:1996 to EN 60835-2-8:1993.

EN 60835-2-10:1993—Methods of measurement for equipment used in digital microwave radio transmission systems—Part 2: Measurements on terrestrial radio-relay systems—Section 10: Overall system performance.

EN 60835-3-5:1995—Methods of measurement for equipment used in digital microwave radio transmission systems—Part 3: Measurements on satellite earth stations—Section 5: Up and down converters.

EN 60835-3-7:1999—Methods of measurement for equipment used in digital microwave radio transmission systems—Part 3: Measurements on satellite earth stations—Section 7: Figure-of-merit of receiving system.

EN 61079-1:1995—Methods of measurement on receivers for satellite broadcast transmissions in the 12 GHz band—Part 1: Radio-frequency measurements on outdoor units.

EN 61079-2:1995—Methods of measurement on receivers for satellite broadcast transmissions in the 12 GHz band—Part 2: Electrical measurements on DBS tuner units.

EN 61079-3:1995—Methods of measurement on receivers for satellite broadcast transmissions in the 12 GHz band—Part 3: Electrical measurements of overall performance of receiver systems comprising an outdoor unit and a DBS tuner unit.

EN 61079-5:1995—Methods of measurement on receivers for satellite broadcast transmissions in the 12 GHz band—Part 5: Electrical measurements on decoder units for MAC/Packet systems.

EN 61108-2:1998—Maritime navigation and radiocommunication equipment and systems—Global navigation satellite systems (GNSS)—Part 2: Global navigation satellite system (GLONASS)—Receiver equipment—Performance standards, methods of testing and required test results.

EN 61114-1:1999—Receiving antennas for satellite broadcast transmissions in the 11–12 GHz band—Part 1: Electrical measurements.

EN 61114-2:1997—Methods of measurement on receiving antennas for satellite broadcast transmission in the 11–12 GHz band—Part 2: Mechanical and environmental tests on individal and collective receiving antennas.

EN 61162-1:2001—Maritime navigation and radiocommunication equipment and systems—Digital interfaces—Part 1: Single talker and multiple listeners.

EN 61162-2:1999—Maritime navigation and radiocommunication equipment and systems—Digital interfaces—Part 2: Single talker and multiple listeners, high-speed transmission.

EN 61603-1:1997—Transmission of audio and/or video and related signals using infrared radiation—Part 1: General.

EN 61603-2:1997—Transmission of audio and/or video and related signals using infrared radiation—Part 2: Transmission systems for audio wideband and related signals.

EN 61603-3:1998—Transmission of audio and/or video and related signals using infrared radiation—Part 3: Transmission systems for audio signals for conference and similar systems.

ETR 069:1993—Radio equipment and systems (RES); High-performance radio local-area network (HIPERLAN); Services and facilities.

ETR 133:1994—Radio equipment and systems (RES); High-performance radio local-area networks (HIPERLAN); System definition.

ETR 152:1996—Transmission and multiplexing (TM); High-bit-rate digital subscriber line (HDSL) transmission system on metallic local line SHDSL core specification and applications for 2048 Kb/sec–based access digital.

ETR 226:1995—Radio equipment and systems (RES); High-performance radio local-area network (HIPERLAN); Architecture for time-bound services (TBS).

ETS 300 001:1997—Attachments to public switched telephone network (PSTN); General technical requirements for equipment connected to an analog subscriber interface in the PSTN.

ETS 300 328:1996—Radio equipment and systems (RES); Wideband transmission systems; Technical characteristics and test conditions for data transmission equipment operating in the 2.4 GHz ISM band and using spread-spectrum modulation techniques; Amendment A1:1997 to ETS 300 328:1996.

Modern Communications Systems

ETS 300 652:1996—Radio equipment and systems (RES); High-performance radio local-area network (HIPERLAN) Type 1; Functional specification; Amendment A1:1997 to ETS 300 652:1996; Amendment A2:1998 to ETS 300 652:1996.

ETS 300 826:1997—Electromagnetic compatibility and radio spectrum matters (ERM); Electromagnetic compatibility (EMC) standard for 2.4 GHz wideband transmission systems and high-performance radio local-area network (HIPERLAN) equipment.

ETS 300 836-1:1997—Broadband radio access networks (BRAN); High-performance radio local-area network (HIPERLAN) Type 1; Conformance testing specification; Part 1: Radio type approval and radio-frequency (RF) conformance test specification.

ETS 300 836-2:1997—Broadband radio access networks (BRAN); High-performance radio local-area network (HIPERLAN) Type 1; Conformance testing specification; Part 2: Protocol implementation conformance statement (PICS) pro forma specification.

ETS 300 836-3:1997—Broadband radio access networks (BRAN); High-performance radio local-area network (HIPERLAN) Type 1; Conformance testing specification; Part 3: Test suite structure and test purposes (TSS&TP) specification.

ETS 300 836-4:1997—Broadband radio access networks (BRAN); High-performance radio local-area network (HIPERLAN) Type 1; Conformance testing specification; Part 4: Abstract test suite (ATS) specification.

HD 466.5 S1:1989—Methods of measurement for radio equipment used in the mobile services—Part 5: Receivers employing single-sideband techniques (R3E, H3E, or J3E).

HD 466.6 S2:1992—Methods of measurement for radio equipment used in the mobile services—Part 6: Selective-calling and data equipment.

HD 466.8 S1:1986—Methods of measurement for radio equipment used in the mobile services—Part 8: Methods of measurement for antennas.

HD 97 S1:1978—Recommended methods of measurement on receivers for frequency-modulation broadcast transmissions.

HD 577 S1:1990—Standardization of interconnections between broadcasting transmitters or transmitter systems and supervisory equipment—Part 1: Interface standards for systems using dedicated interconnections.

Modern Communications Systems

IEEE C63.19:2001—Method of measurement of compatibility between wireless communications devices and hearing aids.

IEEE 187:1990 (R1995)—IEEE standard on radio receivers: Open field method of measurement of spurious radiation from FM and television broadcast receivers.

IEEE 211:1997—IEEE standard definitions of terms for radio wave propagation.

IEEE 269:2002—IEEE standard methods for measuring transmission performance of analog and digital telephone sets.

IEEE 377:1980 (R1997)—IEEE recommended practice for measurement of spurious emission from land-mobile communication transmitters.

IEEE 743:1995—IEEE standard equipment requirements and measurement techniques for analog transmission parameters for telecommunications.

IEEE 802.3:2002 (R2003)—IEEE standard for information technology—Telecommunications and information exchange between systems—Local and metropolitan area networks—Specific requirements—Part 3: Carrier sense multiple access with collision detection (CSMA/CD) access method and physical layer specifications.

IEEE 802.5v:2001—IEEE standard for information technology—Telecommunications and information exchange between systems—Local and metropolitan area networks—Specific requirements—Part 5: Token ring access method and physical layer specifications; Amendment 5: Gigabit token ring operation.

IEEE 802.11:1999 (R2001)—IEEE standard for information technology—Telecommunications and information exchange between systems—Local and metropolitan area networks—Specific requirements—Part 11: Wireless local-area network (LAN) medium-access control (MAC) and physical layer (PHY) specifications.

IEEE 802.11a:1999 (R2001)—IEEE standard for telecommunications and information exchange between systems—LAN/MAN specific requirements—Part 11: Wireless medium-access control (MAC) and physical layer (PHY) specifications: High-speed physical layer in the 5 GHz band.

IEEE 802.11b:1999 (R2001)—IEEE standard for information technology—Telecommunications and information exchange between systems—Local and metropolitan networks—Specific requirements—Part 11: Wireless local-area network (LAN) medium-access control (MAC) and physical layer (PHY) specifications: Higher-speed physical layer (PHY) extension in the 2.4 GHz band.

IEEE 802.15.1:2002—Standard for telecommunications and information exchange between systems—LAN/MAN—Specific requirements—Part 15: Wireless medium-access control (MAC) and physical layer (PHY) specifications for wireless personal-area networks (WPAN).

IEEE 802.16:2001—IEEE local and metropolitan area networks—Part 16: Standard air interface for fixed broadband wireless access systems.

IEEE 820:1984 (R1999)—IEEE standard telephone loop performance characteristics.

IEEE 1007:1991 (R1997)—IEEE standard methods and equipment for measuring the transmission characteristics of pulse-code modulation (PCM) telecommunications circuits and systems.

IEEE 1027:1996—IEEE standard method for measurement of the magnetic field in the vicinity of a telephone receiver.

IEEE 1206:1994—IEEE standard methods for measuring transmission performance of telephone handsets and headsets.

IEEE 1329:1999—IEEE standard method for measuring transmission performance of hands-free telephone sets.

IEEE 1390:1995—IEEE standard for utility telemetry service architecture for switched telephone network.

IEEE 8802-11:1999/IEC 8802-11:1999—Information technology—Telecommunications and information exchange between systems—Local and metropolitan area networks—Specific requirements—Part 11: Wireless LAN medium-access control (MAC) and physical layer (PHY) specifications; Amendment 1:2000 to IEEE 8802-11:1999.

TR 101 031:1999—Broadband radio access networks (BRAN); High-performance radio local-area network (HIPERLAN) Type 2; Requirements and architectures for wireless broadband access.

TR 101 683:2000—Broadband radio access networks (BRAN); HIPERLAN Type 2; System overview.

TR 101 764:2000—Broadband radio access networks (BRAN); Definition of the BRAN domain.

TS 101 475:2000—Broadband radio access networks (BRAN); HIPERLAN Type

2; Physical (PHY) layer.

TS 101 493-2:2000—Broadband radio access networks (BRAN); HIPERLAN Type 2; Packet-based convergence layer; Part 2: Ethernet service specific convergence sublayer (SSCS).

TS 101 761-1:2000—Broadband radio access networks (BRAN); HIPERLAN Type 2; Data link control (DLC) layer; Part 1: Basic data transport functions.

TS 101 761-2:2002—Broadband radio access networks (BRAN); HIPERLAN Type 2; Data link control (DLC) layer; Part 2: Radio link control (RLC) sublayer.

TS 101 762:2000—Broadband radio access networks (BRAN); HIPERLAN Type 2; Network management.

TS 101 763-2:2000—Broadband radio access networks (BRAN); HIPERLAN Type 2; Cell-based convergence layer; Part 2: UNI service specific convergence sublayer (SSCS).

TS 101 811-1-1:2000—Broadband radio access networks (BRAN); HIPERLAN Type 2; Conformance testing for the packet-based convergence layer; Part 1: Common part; Subpart 1: Protocol implementation conformance statement (PICS) pro forma.

TS 101 811-1-2:2001—Broadband radio access networks (BRAN); HIPERLAN Type 2; Conformance testing for the packet-based convergence layer; Part 1: Common part; Subpart 2: Test suite structure and test purposes (TSS&TP) specification.

TS 101 823-1-1:2000—Broadband radio access networks (BRAN); HIPERLAN Type 2; Conformance testing for the data link control (DLC) protocol; Part 1: Basic data transport function; Subpart 1: Protocol implementation conformance statement (PICS) pro forma.

TS 101 823-1-2:2001—Broadband radio access networks (BRAN); HIPERLAN Type 2; Conformance testing for the data link control (DLC) protocol; Part 1: Basic data transport function; Subpart 2: Test suite structure and test purposes (TSS&TP) specification.

TS 101 823-2-1:2001—Broadband radio access networks (BRAN); HIPERLAN Type 2; Conformance testing for the data link control (DLC) protocol; Part 2: Radio link control (RLC) sublayer; Subpart 1: Protocol implementation conformance statement (PICS) pro forma.

TS 101 823-2-2:2001—Broadband radio access networks (BRAN); HIPERLAN Type 2; Conformance testing for the data link control (DLC) protocol; Part 2:

Radio link control (RLC) sublayer; Subpart 2: Test suite structure and test purposes (TSS&TP) specification.

Resources

Code of Federal Regulations

Federal Communications Commission,
http://wireless.fcc.gov/rules.html

National Archives and Records Administration,
http://www.access.gpo.gov/cgi-bin/cfrassemble.cgi?title=200147

IEEE 802 Standards

EEE Standards Association, Get IEEE 802 Program,
http://www.standards.ieee.org/getieee802/

ETSI Telecom Standards
European Telecommunications Standards Institute,
http://www.etsi.org/getastandard/home.htm

ANSI Telecom Standards

American National Standards Institute, http://www.ansi.org

CEN Standards

European Committee for Standardization,
http://www.cenorm.be/catweb/

Global Standards

NSSN: A National Resource for Global Standards,
http://www.nssn.org/

Modern Communications Systems

Appendix B. TCP and UDP ports

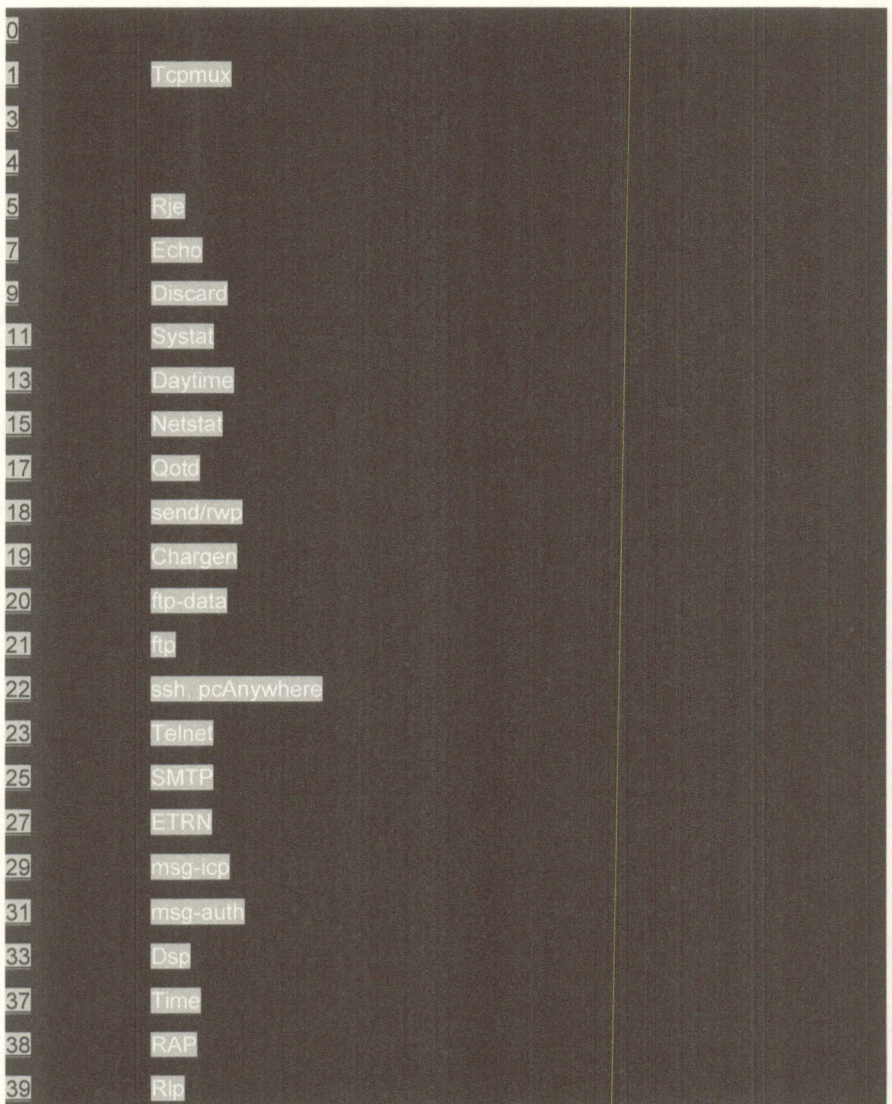

0	
1	Tcpmux
3	
4	
5	Rje
7	Echo
9	Discard
11	Systat
13	Daytime
15	Netstat
17	Qotd
18	send/rwp
19	Chargen
20	ftp-data
21	ftp
22	ssh, pcAnywhere
23	Telnet
25	SMTP
27	ETRN
29	msg-icp
31	msg-auth
33	Dsp
37	Time
38	RAP
39	Rlp

40	
41	
42	nameserv, WINS
43	whois, nickname
49	TACACS, Login Host Protocol
50	RMCP, re-mail-ck
53	DNS
57	MTP
59	NFILE
63	whois++
66	sql*net
67	Bootps
68	bootpd/dhcp
69	Trivial File Transfer Protocol (tftp)
70	Gopher
79	Finger
80	www-http
87	
88	Kerberos, WWW
95	Supdup
96	DIXIE
98	Linuxconf
101	HOSTNAME
102	ISO, X.400, ITOT
105	Cso
106	Poppassd
109	POP2

110	POP3
111	Sun RPC Portmapper
113	Identd/auth
115	Sftp
116	
117	Uucp
118	
119	NNTP
120	CFDP
123	NTP
124	SecureID
129	PWDGEN
133	Statsrv
135	loc-srv/epmap
137	netbios-ns
138	netbios-dgm (UDP)
139	NetBIOS
143	IMAP
144	NewS
150	
152	BFTP
153	SGMP
156	
161	SNMP
175	Vmnet
177	XDMCP
178	NextStep Window Server

179	BGP
180	SLmail admin
199	Smux
210	Z39.50
213	
218	MPP
220	IMAP3
256	
257	
258	
259	ESRO
264	FW1_topo
311	Apple WebAdmin
350	MATIP type A
351	MATIP type B
360	
363	RSVP tunnel
366	ODMR (On-Demand Mail Relay)
371	
387	AURP (AppleTalk Update-Based Routing Protocol)
389	LDAP
407	Timbuktu
427	
434	Mobile IP
443	Ssl
444	snpp, Simple Network Paging Protocol
445	SMB

458	QuickTime TV/Conferencing
468	Photuris
475	
500	ISAKMP, Pluto
511	
512	biff, rexec
513	who, rlogin
514	syslog, rsh
515	lp, lpr, line printer
517	Talk
520	RIP (Routing Information Protocol)
521	RIPng
522	ULS
531	IRC
543	KLogin, AppleShare over IP
545	QuickTime
548	AFP
554	Real Time Streaming Protocol
555	phAse Zero
563	NNTP over SSL
575	VEMMI
581	Bundle Discovery Protocol
593	MS-RPC
608	SIFT/UFT
626	Apple ASIA
631	IPP (Internet Printing Protocol)
635	Mountd

636	Sldap
642	EMSD
648	RRP (NSI Registry Registrar Protocol)
655	Tinc
660	Apple MacOS Server Admin
666	Doom
674	ACAP
687	AppleShare IP Registry
700	Buddyphone
705	AgentX for SNMP
901	swat, realsecure
993	s-imap
995	s-pop
999	
1024	
1025	
1050	
1062	Veracity
1080	SOCKS
1085	WebObjects
1100	
1105	
1114	
1227	DNS2Go
1234	
1243	SubSeven
1338	Millennium Worm

1352	Lotus Notes
1381	Apple Network License Manager
1417	Timbuktu
1418	Timbuktu
1419	Timbuktu
1420	
1433	Microsoft SQL Server
1434	Microsoft SQL Monitor
1477	
1478	
1490	
1494	Citrix ICA Protocol
1498	
1500	
1503	T.120
1521	Oracle SQL
1522	
1524	
1525	prospero
1526	prospero
1527	Tlisrv
1529	
1547	
1604	Citrix ICA, MS Terminal Server
1645	RADIUS Authentication
1646	RADIUS Accounting
1680	Carbon Copy

Modern Communications Systems

1701	L2TP/LSF
1717	Convoy
1720	H.323/Q.931
1723	PPTP control port
1731	
1755	Windows Media .asf
1758	TFTP multicast
1761	
1762	
1808	
1812	RADIUS server
1813	RADIUS accounting
1818	ETFTP
1968	
1973	DLSw DCAP/DRAP
1975	
1978	
1979	
1985	HSRP
1999	Cisco AUTH
2000	
2001	Glimpse
2005	
2010	
2023	
2048	
2049	NFS

2064	distributed.net
2065	DLSw
2066	DLSw
2080	
2106	MZAP
2140	DeepThroat
2301	Compaq Insight Management Web Agents
2327	Netscape Conference
2336	Apple UG Control
2345	
2427	MGCP gateway
2504	WLBS
2535	MADCAP
2543	Sip
2565	
2592	Netrek
2727	MGCP call agent
2766	
2628	DICT
2998	ISS Real Secure Console Service Port
3000	Firstclass
3001	
3031	Apple AgentVU
3052	
3128	Squid
3130	ICP
3150	DeepThroat

3264	Ccmail
3283	Apple NetAssitant
3288	COPS
3305	ODETTE
3306	mySQL
3352	
3389	RDP Protocol (Terminal Server)
3520	
3521	Netrek
3879	
4000	icq, command-n-conquer
4045	
4144	
4242	
4321	Rwhois
4333	mSQL
4444	
47017	
4827	HTCP
5000	
5001	
5002	
5004	RTP
5005	RTP
5010	Yahoo! Messenger
5050	
5060	SIP

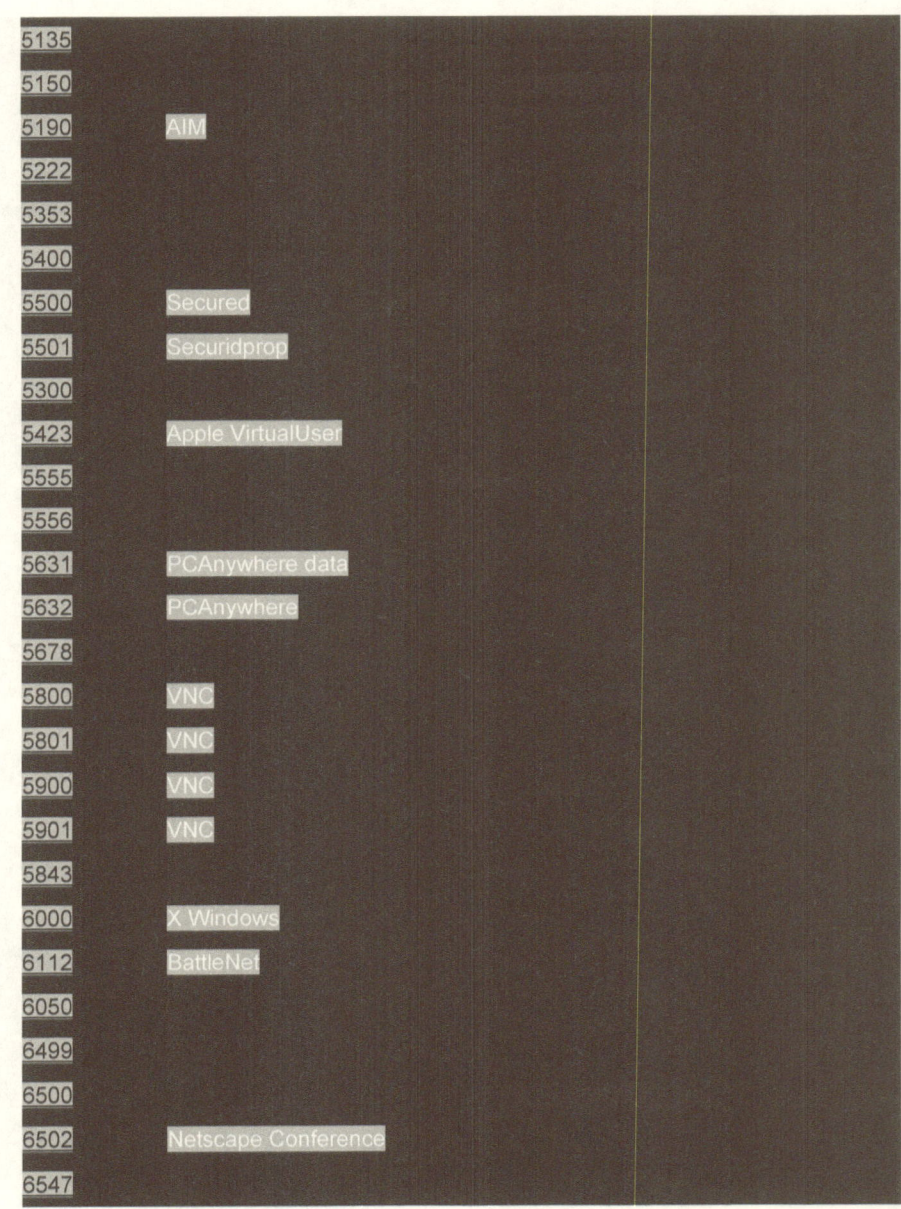

5135	
5150	
5190	AIM
5222	
5353	
5400	
5500	Secured
5501	Securidprop
5300	
5423	Apple VirtualUser
5555	
5556	
5631	PCAnywhere data
5632	PCAnywhere
5678	
5800	VNC
5801	VNC
5900	VNC
5901	VNC
5843	
6000	X Windows
6112	BattleNet
6050	
6499	
6500	
6502	Netscape Conference
6547	

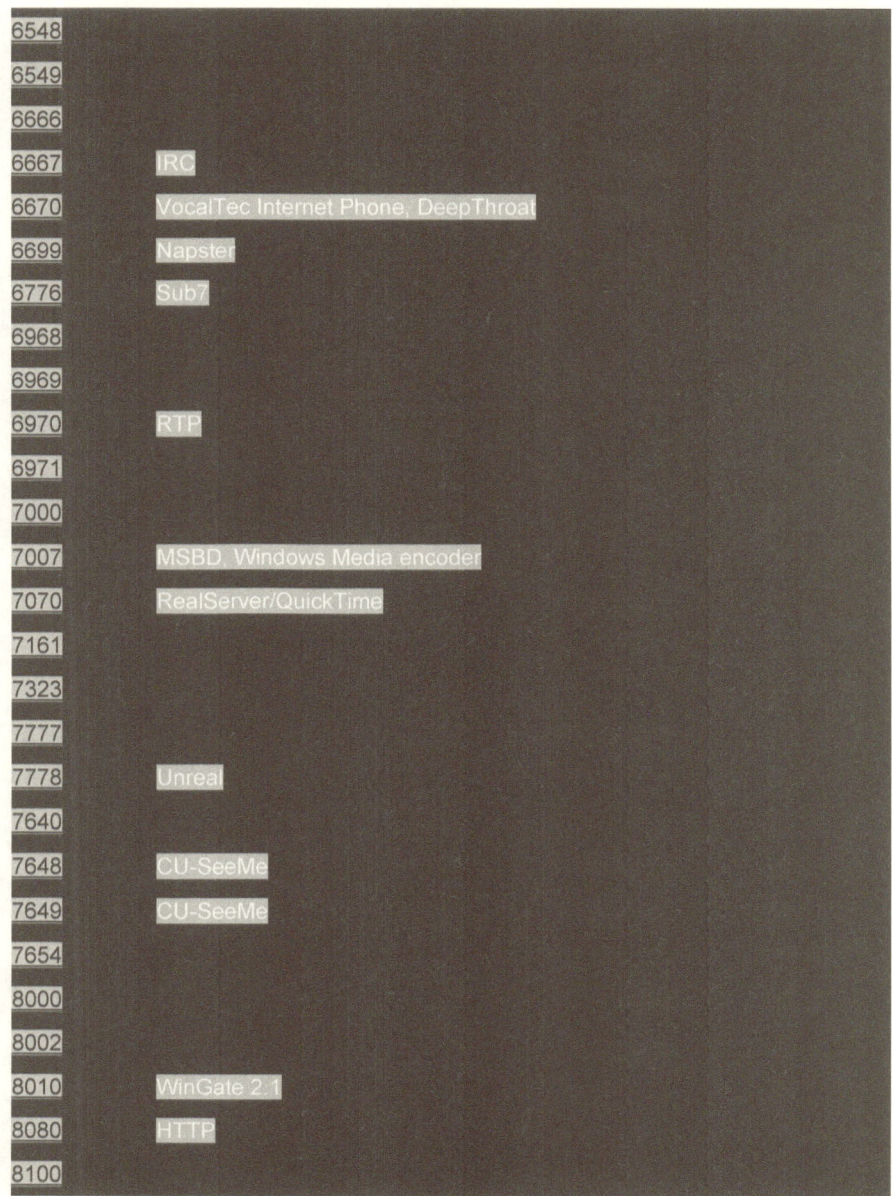

6548	
6549	
6666	
6667	IRC
6670	VocalTec Internet Phone, DeepThroat
6699	Napster
6776	Sub7
6968	
6969	
6970	RTP
6971	
7000	
7007	MSBD, Windows Media encoder
7070	RealServer/QuickTime
7161	
7323	
7777	
7778	Unreal
7640	
7648	CU-SeeMe
7649	CU-SeeMe
7654	
8000	
8002	
8010	WinGate 2.1
8080	HTTP
8100	

211

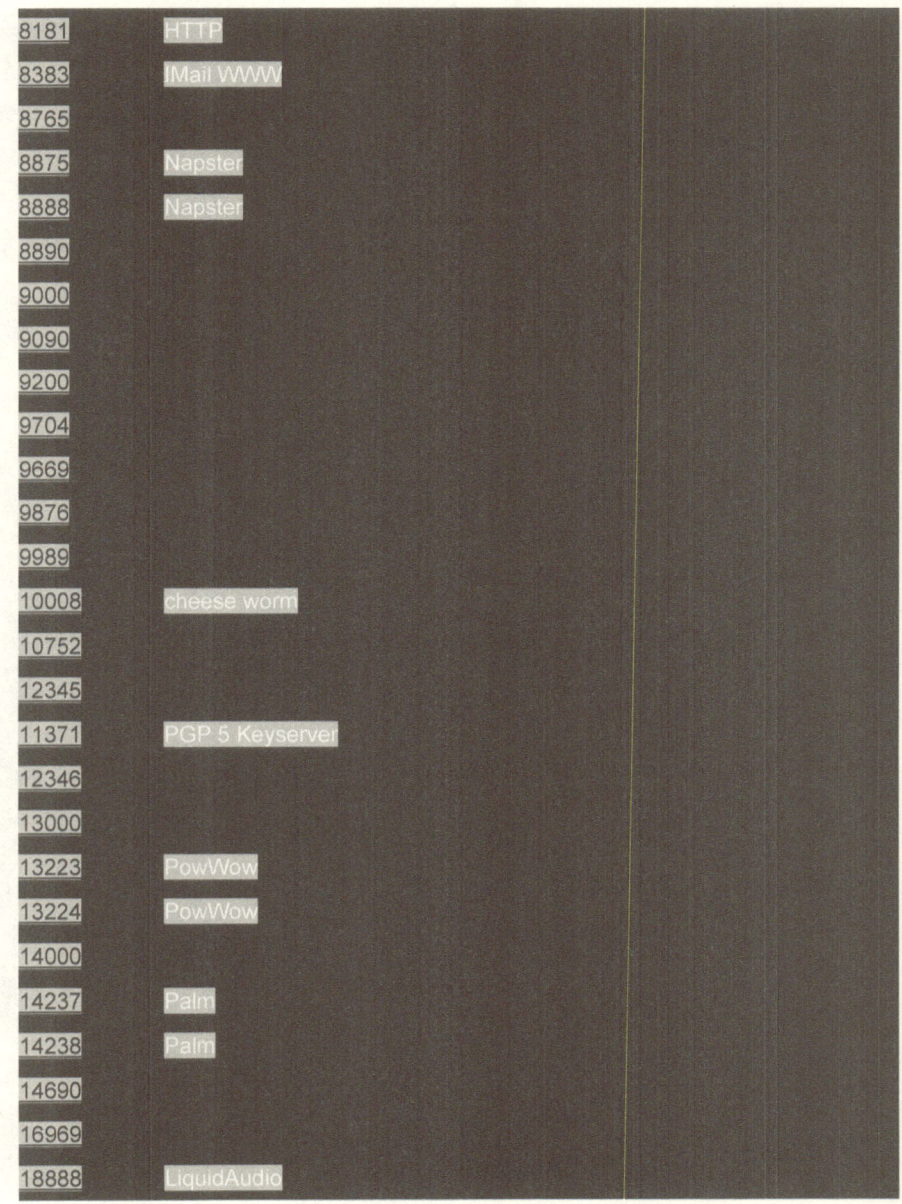

Port	Service
8181	HTTP
8383	IMail WWW
8765	
8875	Napster
8888	Napster
8890	
9000	
9090	
9200	
9704	
9669	
9876	
9989	
10008	cheese worm
10752	
12345	
11371	PGP 5 Keyserver
12346	
13000	
13223	PowWow
13224	PowWow
14000	
14237	Palm
14238	Palm
14690	
16969	
18888	LiquidAudio

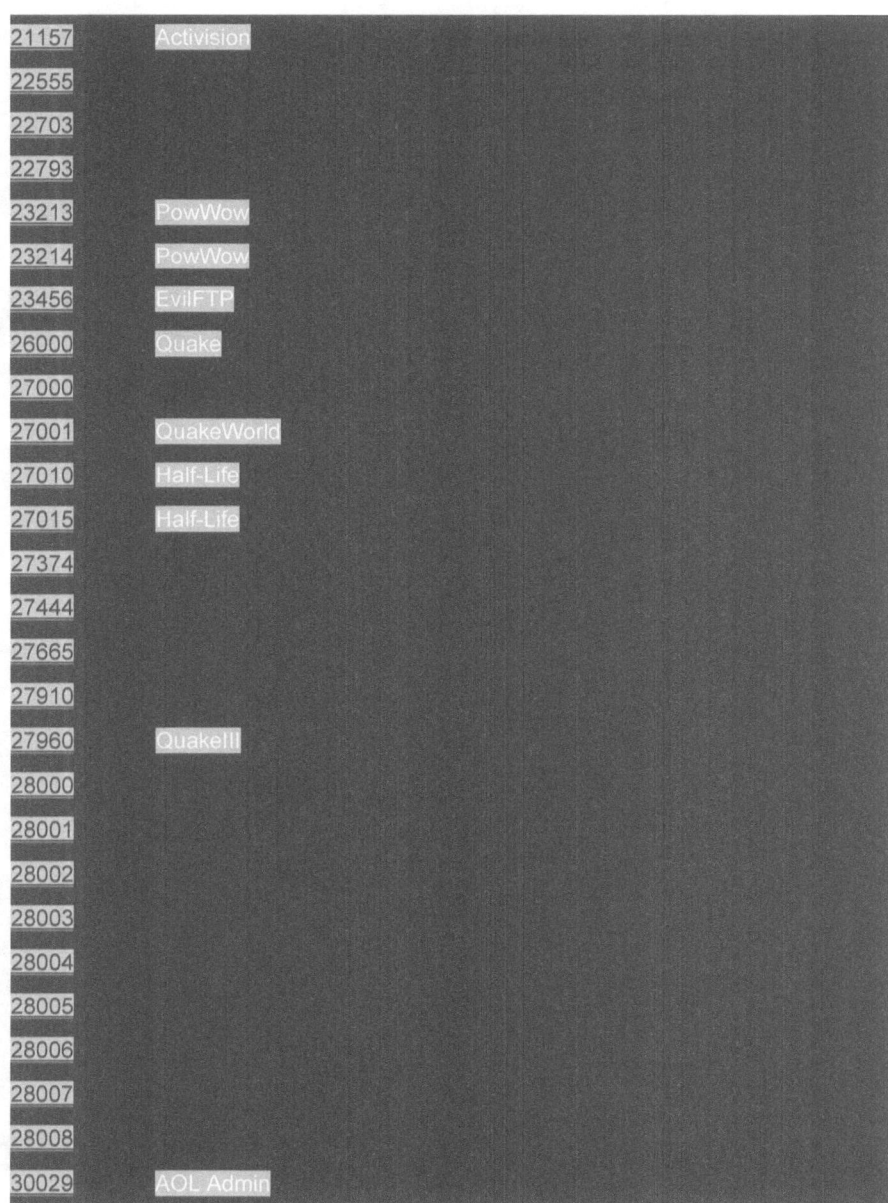

21157	Activision
22555	
22703	
22793	
23213	PowWow
23214	PowWow
23456	EvilFTP
26000	Quake
27000	
27001	QuakeWorld
27010	Half-Life
27015	Half-Life
27374	
27444	
27665	
27910	
27960	QuakeIII
28000	
28001	
28002	
28003	
28004	
28005	
28006	
28007	
28008	
30029	AOL Admin

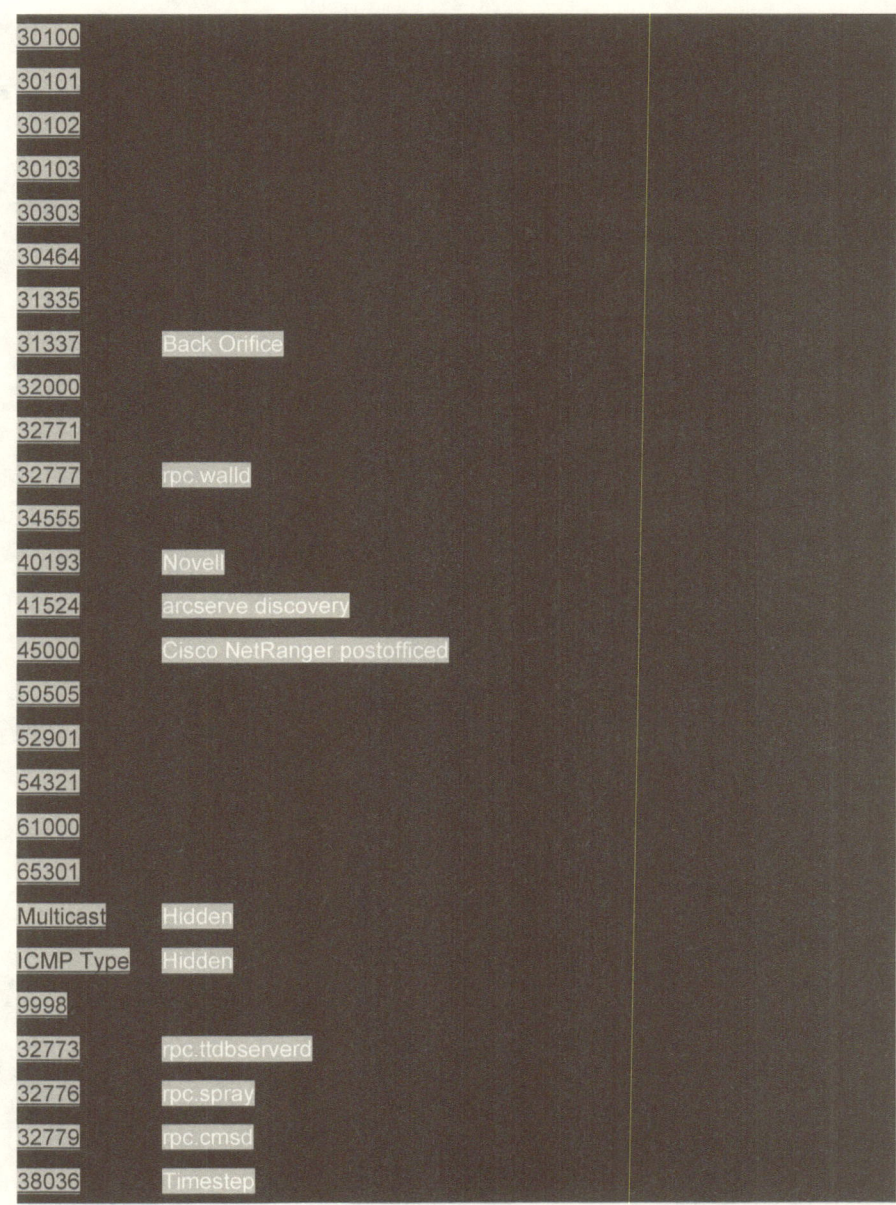

30100	
30101	
30102	
30103	
30303	
30464	
31335	
31337	Back Orifice
32000	
32771	
32777	rpc.walld
34555	
40193	Novell
41524	arcserve discovery
45000	Cisco NetRanger postofficed
50505	
52901	
54321	
61000	
65301	
Multicast	Hidden
ICMP Type	Hidden
9998	
32773	rpc.ttdbserverd
32776	rpc.spray
32779	rpc.cmsd
38036	Timestep

Appendix C. Port Assignments:

Keyword	Decimal	Description	References
-------	-------	-----------	----------
	0/tcp	Reserved	
	0/udp	Reserved	
#		Jon Postel <postel@isi.edu>	
tcpmux	1/tcp	TCP Port Service Multiplexer	
tcpmux	1/udp	TCP Port Service Multiplexer	
#		Mark Lottor <MKL@nisc.sri.com>	
compressnet	2/tcp	Management Utility	
compressnet	2/udp	Management Utility	
compressnet	3/tcp	Compression Process	
compressnet	3/udp	Compression Process	
#		Bernie Volz <VOLZ@PROCESS.COM>	
#	4/tcp	Unassigned	
#	4/udp	Unassigned	
rje	5/tcp	Remote Job Entry	
rje	5/udp	Remote Job Entry	
#		Jon Postel <postel@isi.edu>	
#	6/tcp	Unassigned	
#	6/udp	Unassigned	
echo	7/tcp	Echo	
echo	7/udp	Echo	
#		Jon Postel <postel@isi.edu>	
#	8/tcp	Unassigned	
#	8/udp	Unassigned	
discard	9/tcp	Discard	
discard	9/udp	Discard	
#		Jon Postel <postel@isi.edu>	
#	10/tcp	Unassigned	
#	10/udp	Unassigned	
systat	11/tcp	Active Users	
systat	11/udp	Active Users	
#		Jon Postel <postel@isi.edu>	
#	12/tcp	Unassigned	
#	12/udp	Unassigned	
daytime	13/tcp	Daytime (RFC 867)	
daytime	13/udp	Daytime (RFC 867)	
#		Jon Postel <postel@isi.edu>	
#	14/tcp	Unassigned	
#	14/udp	Unassigned	
#	15/tcp	Unassigned [was netstat	
#	15/udp	Unassigned	
#	16/tcp	Unassigned	

#	16/udp	Unassigned
qotd	17/tcp	Quote of the Day
qotd	17/udp	Quote of the Day
#	Jon Postel <postel@isi.edu>	
msp	18/tcp	Message Send Protocol
msp	18/udp	Message Send Protocol
#	Rina Nethaniel <---none--->	
chargen	19/tcp	Character Generator
chargen	19/udp	Character Generator
ftp-data	20/tcp	File Transfer [Default Data]
ftp-data	20/udp	File Transfer [Default Data]
ftp	21/tcp	File Transfer [Control]
ftp	21/udp	File Transfer [Control]
#	Jon Postel <postel@isi.edu>	
ssh	22/tcp	SSH Remote Login Protocol
ssh	22/udp	SSH Remote Login Protocol
#	Tatu Ylonen <ylo@cs.hut.fi>	
telnet	23/tcp	Telnet
telnet	23/udp	Telnet
#	Jon Postel <postel@isi.edu>	
	24/tcp	any private mail system
	24/udp	any private mail system
#	Rick Adams <rick@UUNET.UU.NET>	
smtp	25/tcp	Simple Mail Transfer
smtp	25/udp	Simple Mail Transfer
#	Jon Postel <postel@isi.edu>	
#	26/tcp	Unassigned
#	26/udp	Unassigned
nsw-fe	27/tcp	NSW User System FE
nsw-fe	27/udp	NSW User System FE
#	Robert Thomas <BThomas@F.BBN.COM>	
#	28/tcp	Unassigned
#	28/udp	Unassigned
msg-icp	29/tcp	MSG ICP
msg-icp	29/udp	MSG ICP
#	Robert Thomas <BThomas@F.BBN.COM>	
#	30/tcp	Unassigned
#	30/udp	Unassigned
msg-auth	31/tcp	MSG Authentication
msg-auth	31/udp	MSG Authentication
#	Robert Thomas <BThomas@F.BBN.COM>	
#	32/tcp	Unassigned
#	32/udp	Unassigned
dsp	33/tcp	Display Support Protocol
dsp	33/udp	Display Support Protocol
#	Ed Cain <cain@edn-unix.dca.mil>	

#	34/tcp	Unassigned
#	34/udp	Unassigned
	35/tcp	any private printer server
	35/udp	any private printer server
#	Jon Postel <postel@isi.edu>	
#	36/tcp	Unassigned
#	36/udp	Unassigned
time	37/tcp	Time
time	37/udp	Time
#	Jon Postel <postel@isi.edu>	
rap	38/tcp	Route Access Protocol
rap	38/udp	Route Access Protocol
#	Robert Ullmann <ariel@world.std.com>	
rlp	39/tcp	Resource Location Protocol
rlp	39/udp	Resource Location Protocol
#	Mike Accetta <MIKE.ACCETTA@CMU-CS-A.EDU>	
#	40/tcp	Unassigned
#	40/udp	Unassigned
graphics	41/tcp	Graphics
graphics	41/udp	Graphics
name	42/tcp	Host Name Server
name	42/udp	Host Name Server
nameserver	42/tcp	Host Name Server
nameserver	42/udp	Host Name Server
nicname	43/tcp	Who Is
nicname	43/udp	Who Is
mpm-flags	44/tcp	MPM FLAGS Protocol
mpm-flags	44/udp	MPM FLAGS Protocol
mpm	45/tcp	Message Processing Module [recv]
mpm	45/udp	Message Processing Module [recv]
mpm-snd	46/tcp	MPM [default send]
mpm-snd	46/udp	MPM [default send]
#	Jon Postel <postel@isi.edu>	
ni-ftp	47/tcp	NI FTP
ni-ftp	47/udp	NI FTP
#	Steve Kille <S.Kille@isode.com>	
auditd	48/tcp	Digital Audit Daemon
auditd	48/udp	Digital Audit Daemon
#	Larry Scott <scott@zk3.dec.com>	
tacacs	49/tcp	Login Host Protocol (TACACS)
tacacs	49/udp	Login Host Protocol (TACACS)
#	Pieter Ditmars <pditmars@BBN.COM>	
re-mail-ck	50/tcp	Remote Mail Checking Protocol
re-mail-ck	50/udp	Remote Mail Checking Protocol
#	Steve Dorner <s-dorner@UIUC.EDU>	
la-maint	51/tcp	IMP Logical Address Maintenance

la-maint	51/udp	IMP Logical Address Maintenance
#	Andy Malis <malis_a@timeplex.com>	
xns-time	52/tcp	XNS Time Protocol
xns-time	52/udp	XNS Time Protocol
#	Susie Armstrong <Armstrong.wbst128@XEROX>	
domain	53/tcp	Domain Name Server
domain	53/udp	Domain Name Server
#	Paul Mockapetris <PVM@ISI.EDU>	
xns-ch	54/tcp	XNS Clearinghouse
xns-ch	54/udp	XNS Clearinghouse
#	Susie Armstrong <Armstrong.wbst128@XEROX>	
isi-gl	55/tcp	ISI Graphics Language
isi-gl	55/udp	ISI Graphics Language
xns-auth	56/tcp	XNS Authentication
xns-auth	56/udp	XNS Authentication
#	Susie Armstrong <Armstrong.wbst128@XEROX>	
	57/tcp	any private terminal access
	57/udp	any private terminal access
#	Jon Postel <postel@isi.edu>	
xns-mail	58/tcp	XNS Mail
xns-mail	58/udp	XNS Mail
#	Susie Armstrong <Armstrong.wbst128@XEROX>	
	59/tcp	any private file service
	59/udp	any private file service
#	Jon Postel <postel@isi.edu>	
	60/tcp	Unassigned
	60/udp	Unassigned
ni-mail	61/tcp	NI MAIL
ni-mail	61/udp	NI MAIL
#	Steve Kille <S.Kille@isode.com>	
acas	62/tcp	ACA Services
acas	62/udp	ACA Services
#	E. Wald <ewald@via.enet.dec.com>	
whois++	63/tcp	whois++
whois++	63/udp	whois++
#	Rickard Schoultz <schoultz@sunet.se>	
covia	64/tcp	Communications Integrator (CI)
covia	64/udp	Communications Integrator (CI)
#	Dan Smith <dan.smith@den.galileo.com>	
tacacs-ds	65/tcp	TACACS-Database Service
tacacs-ds	65/udp	TACACS-Database Service
#	Kathy Huber <khuber@bbn.com>	
sql*net	66/tcp	Oracle SQL*NET
sql*net	66/udp	Oracle SQL*NET
#	Jack Haverty <jhaverty@ORACLE.COM>	
bootps	67/tcp	Bootstrap Protocol Server

bootps	67/udp	Bootstrap Protocol Server
bootpc	68/tcp	Bootstrap Protocol Client
bootpc	68/udp	Bootstrap Protocol Client
#	Bill Croft <Croft@SUMEX-AIM.STANFORD.EDU>	
tftp	69/tcp	Trivial File Transfer
tftp	69/udp	Trivial File Transfer
#	David Clark <ddc@LCS.MIT.EDU>	
gopher	70/tcp	Gopher
gopher	70/udp	Gopher
#	Mark McCahill <mpm@boombox.micro.umn.edu>	
netrjs-1	71/tcp	Remote Job Service
netrjs-1	71/udp	Remote Job Service
netrjs-2	72/tcp	Remote Job Service
netrjs-2	72/udp	Remote Job Service
netrjs-3	73/tcp	Remote Job Service
netrjs-3	73/udp	Remote Job Service
netrjs-4	74/tcp	Remote Job Service
netrjs-4	74/udp	Remote Job Service
#	Bob Braden <Braden@ISI.EDU>	
	75/tcp	any private dial out service
	75/udp	any private dial out service
#	Jon Postel <postel@isi.edu>	
deos	76/tcp	Distributed External Object Store
deos	76/udp	Distributed External Object Store
#	Robert Ullmann <ariel@world.std.com>	
	77/tcp	any private RJE service
	77/udp	any private RJE service
#	Jon Postel <postel@isi.edu>	
vettcp	78/tcp	vettcp
vettcp	78/udp	vettcp
#	Christopher Leong <leong@kolmod.mlo.dec.com>	
finger	79/tcp	Finger
finger	79/udp	Finger
#	David Zimmerman <dpz@RUTGERS.EDU>	
http	80/tcp	World Wide Web HTTP
http	80/udp	World Wide Web HTTP
www	80/tcp	World Wide Web HTTP
www	80/udp	World Wide Web HTTP
www-http	80/tcp	World Wide Web HTTP
www-http	80/udp	World Wide Web HTTP
#	Tim Berners-Lee <timbl@W3.org>	
hosts2-ns	81/tcp	HOSTS2 Name Server
hosts2-ns	81/udp	HOSTS2 Name Server
#	Earl Killian <EAK@MORDOR.S1.GOV>	
xfer	82/tcp	XFER Utility
xfer	82/udp	XFER Utility

#	Thomas M. Smith <Thomas.M.Smith@lmco.com>	
mit-ml-dev	83/tcp	MIT ML Device
mit-ml-dev	83/udp	MIT ML Device
#	David Reed <--none--->	
ctf	84/tcp	Common Trace Facility
ctf	84/udp	Common Trace Facility
#	Hugh Thomas <thomas@oils.enet.dec.com>	
mit-ml-dev	85/tcp	MIT ML Device
mit-ml-dev	85/udp	MIT ML Device
#	David Reed <--none--->	
mfcobol	86/tcp	Micro Focus Cobol
mfcobol	86/udp	Micro Focus Cobol
#	Simon Edwards <--none--->	
	87/tcp	any private terminal link
	87/udp	any private terminal link
#	Jon Postel <postel@isi.edu>	
kerberos	88/tcp	Kerberos
kerberos	88/udp	Kerberos
#	B. Clifford Neuman <bcn@isi.edu>	
su-mit-tg	89/tcp	SU/MIT Telnet Gateway
su-mit-tg	89/udp	SU/MIT Telnet Gateway
#	Mark Crispin <MRC@PANDA.COM>	
########## PORT 90 also being used unofficially by Pointcast #########		
dnsix	90/tcp	DNSIX Securit Attribute Token Map
dnsix	90/udp	DNSIX Securit Attribute Token Map
#	Charles Watt <watt@sware.com>	
mit-dov	91/tcp	MIT Dover Spooler
mit-dov	91/udp	MIT Dover Spooler
#	Eliot Moss <EBM@XX.LCS.MIT.EDU>	
npp	92/tcp	Network Printing Protocol
npp	92/udp	Network Printing Protocol
#	Louis Mamakos <louie@sayshell.umd.edu>	
dcp	93/tcp	Device Control Protocol
dcp	93/udp	Device Control Protocol
#	Daniel Tappan <Tappan@BBN.COM>	
objcall	94/tcp	Tivoli Object Dispatcher
objcall	94/udp	Tivoli Object Dispatcher
#	Tom Bereiter <--none--->	
supdup	95/tcp	SUPDUP
supdup	95/udp	SUPDUP
#	Mark Crispin <MRC@PANDA.COM>	
dixie	96/tcp	DIXIE Protocol Specification
dixie	96/udp	DIXIE Protocol Specification
#	Tim Howes <Tim.Howes@terminator.cc.umich.edu>	
swift-rvf	97/tcp	Swift Remote Virtural File Protocol
swift-rvf	97/udp	Swift Remote Virtural File Protocol

Modern Communications Systems

```
#                    Maurice R. Turcotte
#                    <mailrus!uflorida!rm1!dnmrt%rmatl@uunet.UU.NET>
tacnews              98/tcp    TAC News
tacnews              98/udp    TAC News
#                    Jon Postel <postel@isi.edu>
metagram             99/tcp    Metagram Relay
metagram             99/udp    Metagram Relay
#                    Geoff Goodfellow <Geoff@FERNWOOD.MPK.CA.US>
newacct              100/tcp   [unauthorized use]
hostname             101/tcp   NIC Host Name Server
hostname             101/udp   NIC Host Name Server
#                    Jon Postel <postel@isi.edu>
iso-tsap             102/tcp   ISO-TSAP Class 0
iso-tsap             102/udp   ISO-TSAP Class 0
#                    Marshall Rose <mrose@dbc.mtview.ca.us>
gppitnp              103/tcp   Genesis Point-to-Point Trans Net
gppitnp              103/udp   Genesis Point-to-Point Trans Net
acr-nema             104/tcp   ACR-NEMA Digital Imag. & Comm. 300
acr-nema             104/udp   ACR-NEMA Digital Imag. & Comm. 300
#                    Patrick McNamee <--none--->
cso                  105/tcp   CCSO name server protocol
cso                  105/udp   CCSO name server protocol
#                    Martin Hamilton <martin@mrrl.lut.as.uk>
csnet-ns             105/tcp   Mailbox Name Nameserver
csnet-ns             105/udp   Mailbox Name Nameserver
#                    Marvin Solomon <solomon@CS.WISC.EDU>
3com-tsmux           106/tcp   3COM-TSMUX
3com-tsmux           106/udp   3COM-TSMUX
#                    Jeremy Siegel <jzs@NSD.3Com.COM>
##########           106       Unauthorized use by insecure poppassd protocol
rtelnet              107/tcp   Remote Telnet Service
rtelnet              107/udp   Remote Telnet Service
#                    Jon Postel <postel@isi.edu>
snagas               108/tcp   SNA Gateway Access Server
snagas               108/udp   SNA Gateway Access Server
#                    Kevin Murphy <murphy@sevens.lkg.dec.com>
pop2                 109/tcp   Post Office Protocol - Version 2
pop2                 109/udp   Post Office Protocol - Version 2
#                    Joyce K. Reynolds <jkrey@isi.edu>
pop3                 110/tcp   Post Office Protocol - Version 3
pop3                 110/udp   Post Office Protocol - Version 3
#                    Marshall Rose <mrose@dbc.mtview.ca.us>
sunrpc               111/tcp   SUN Remote Procedure Call
sunrpc               111/udp   SUN Remote Procedure Call
#                    Chuck McManis <cmcmanis@freegate.net>
mcidas               112/tcp   McIDAS Data Transmission Protocol
```

mcidas	112/udp	McIDAS Data Transmission Protocol
#	Glenn Davis <support@unidata.ucar.edu>	
ident	113/tcp	
auth	113/tcp	Authentication Service
auth	113/udp	Authentication Service
#	Mike St. Johns <stjohns@arpa.mil>	
audionews	114/tcp	Audio News Multicast
audionews	114/udp	Audio News Multicast
#	Martin Forssen <maf@dtek.chalmers.se>	
sftp	115/tcp	Simple File Transfer Protocol
sftp	115/udp	Simple File Transfer Protocol
#	Mark Lottor <MKL@nisc.sri.com>	
ansanotify	116/tcp	ANSA REX Notify
ansanotify	116/udp	ANSA REX Notify
#	Nicola J. Howarth <njh@ansa.co.uk>	
uucp-path	117/tcp	UUCP Path Service
uucp-path	117/udp	UUCP Path Service
sqlserv	118/tcp	SQL Services
sqlserv	118/udp	SQL Services
#	Larry Barnes <barnes@broke.enet.dec.com>	
nntp	119/tcp	Network News Transfer Protocol
nntp	119/udp	Network News Transfer Protocol
#	Phil Lapsley <phil@UCBARPA.BERKELEY.EDU>	
cfdptkt	120/tcp	CFDPTKT
cfdptkt	120/udp	CFDPTKT
#	John Ioannidis <ji@close.cs.columbia.ed>	
erpc	121/tcp	Encore Expedited Remote Pro.Call
erpc	121/udp	Encore Expedited Remote Pro.Call
#	Jack O'Neil <---none--->	
smakynet	122/tcp	SMAKYNET
smakynet	122/udp	SMAKYNET
#	Pierre Arnaud <pierre.arnaud@iname.com>	
ntp	123/tcp	Network Time Protocol
ntp	123/udp	Network Time Protocol
#	Dave Mills <Mills@HUEY.UDEL.EDU>	
ansatrader	124/tcp	ANSA REX Trader
ansatrader	124/udp	ANSA REX Trader
#	Nicola J. Howarth <njh@ansa.co.uk>	
locus-map	125/tcp	Locus PC-Interface Net Map Ser
locus-map	125/udp	Locus PC-Interface Net Map Ser
#	Eric Peterson <lcc.eric@SEAS.UCLA.EDU>	
nxedit	126/tcp	NXEdit
nxedit	126/udp	NXEdit
#	Don Payette <Don.Payette@unisys.com>	

###########Port 126 Previously assigned to application below#######

#unitary	126/tcp	Unisys Unitary Login

```
#unitary              126/udp   Unisys Unitary Login
#                     <feil@kronos.nisd.cam.unisys.com>
###########Port 126 Previously assigned to application above######
locus-con             127/tcp   Locus PC-Interface Conn Server
locus-con             127/udp   Locus PC-Interface Conn Server
#                     Eric Peterson <lcc.eric@SEAS.UCLA.EDU>
gss-xlicen            128/tcp   GSS X License Verification
gss-xlicen            128/udp   GSS X License Verification
#                     John Light <johnl@gssc.gss.com>
pwdgen                129/tcp   Password Generator Protocol
pwdgen                129/udp   Password Generator Protocol
#                     Frank J. Wacho <WANCHO@WSMR-SIMTEL20.ARMY.MIL>
cisco-fna             130/tcp   cisco FNATIVE
cisco-fna             130/udp   cisco FNATIVE
cisco-tna             131/tcp   cisco TNATIVE
cisco-tna             131/udp   cisco TNATIVE
cisco-sys             132/tcp   cisco SYSMAINT
cisco-sys             132/udp   cisco SYSMAINT
statsrv               133/tcp   Statistics Service
statsrv               133/udp   Statistics Service
#                     Dave Mills <Mills@HUEY.UDEL.EDU>
ingres-net            134/tcp   INGRES-NET Service
ingres-net            134/udp   INGRES-NET Service
#                     Mike Berrow <---none--->
epmap                 135/tcp   DCE endpoint resolution
epmap                 135/udp   DCE endpoint resolution
#                     Joe Pato <pato@apollo.hp.com>
profile               136/tcp   PROFILE Naming System
profile               136/udp   PROFILE Naming System
#                     Larry Peterson <llp@ARIZONA.EDU>
netbios-ns            137/tcp   NETBIOS Name Service
netbios-ns            137/udp   NETBIOS Name Service
netbios-dgm           138/tcp   NETBIOS Datagram Service
netbios-dgm           138/udp   NETBIOS Datagram Service
netbios-ssn           139/tcp   NETBIOS Session Service
netbios-ssn           139/udp   NETBIOS Session Service
#                     Jon Postel <postel@isi.edu>
emfis-data            140/tcp   EMFIS Data Service
emfis-data            140/udp   EMFIS Data Service
emfis-cntl            141/tcp   EMFIS Control Service
emfis-cntl            141/udp   EMFIS Control Service
#                     Gerd Beling <GBELING@ISI.EDU>
bl-idm                142/tcp   Britton-Lee IDM
bl-idm                142/udp   Britton-Lee IDM
#                     Susie Snitzer <---none--->
imap                  143/tcp   Internet Message Access Protocol
```

Modern Communications Systems

imap	143/udp	Internet Message Access Protocol
#	Mark Crispin <MRC@CAC.Washington.EDU>	
uma	144/tcp	Universal Management Architecture
uma	144/udp	Universal Management Architecture
#	Jay Whitney <jw@powercenter.com>	
uaac	145/tcp	UAAC Protocol
uaac	145/udp	UAAC Protocol
#	David A. Gomberg <gomberg@GATEWAY.MITRE.ORG>	
iso-tp0	146/tcp	ISO-IP0
iso-tp0	146/udp	ISO-IP0
iso-ip	147/tcp	ISO-IP
iso-ip	147/udp	ISO-IP
#	Marshall Rose <mrose@dbc.mtview.ca.us>	
jargon	148/tcp	Jargon
jargon	148/udp	Jargon
#	Bill Weinman <wew@bearnet.com>	
aed-512	149/tcp	AED 512 Emulation Service
aed-512	149/udp	AED 512 Emulation Service
#	Albert G. Broscius <broscius@DSL.CIS.UPENN.EDU>	
sql-net	150/tcp	SQL-NET
sql-net	150/udp	SQL-NET
#	Martin Picard <<---none--->	
hems	151/tcp	HEMS
hems	151/udp	HEMS
bftp	152/tcp	Background File Transfer Program
bftp	152/udp	Background File Transfer Program
#	Annette DeSchon <DESCHON@ISI.EDU>	
sgmp	153/tcp	SGMP
sgmp	153/udp	SGMP
#	Marty Schoffstahl <schoff@NISC.NYSER.NET>	
netsc-prod	154/tcp	NETSC
netsc-prod	154/udp	NETSC
netsc-dev	155/tcp	NETSC
netsc-dev	155/udp	NETSC
#	Sergio Heker <heker@JVNCC.CSC.ORG>	
sqlsrv	156/tcp	SQL Service
sqlsrv	156/udp	SQL Service
#	Craig Rogers <Rogers@ISI.EDU>	
knet-cmp	157/tcp	KNET/VM Command/Message Protocol
knet-cmp	157/udp	KNET/VM Command/Message Protocol
#	Gary S. Malkin <GMALKIN@XYLOGICS.COM>	
pcmail-srv	158/tcp	PCMail Server
pcmail-srv	158/udp	PCMail Server
#	Mark L. Lambert <markl@PTT.LCS.MIT.EDU>	
nss-routing	159/tcp	NSS-Routing
nss-routing	159/udp	NSS-Routing

```
#                    Yakov Rekhter <Yakov@IBM.COM>
sgmp-traps           160/tcp   SGMP-TRAPS
sgmp-traps           160/udp   SGMP-TRAPS
#                    Marty Schoffstahl <schoff@NISC.NYSER.NET>
snmp                 161/tcp   SNMP
snmp                 161/udp   SNMP
snmptrap             162/tcp   SNMPTRAP
snmptrap             162/udp   SNMPTRAP
#                    Marshall Rose <mrose@dbc.mtview.ca.us>
cmip-man             163/tcp   CMIP/TCP Manager
cmip-man             163/udp   CMIP/TCP Manager
cmip-agent           164/tcp   CMIP/TCP Agent
cmip-agent           164/udp   CMIP/TCP Agent
#                    Amatzia Ben-Artzi <---none--->
xns-courier          165/tcp   Xerox
xns-courier          165/udp   Xerox
#                    Susie Armstrong <Armstrong.wbst128@XEROX.COM>
s-net                166/tcp   Sirius Systems
s-net                166/udp   Sirius Systems
#                    Brian Lloyd <brian@lloyd.com>
namp                 167/tcp   NAMP
namp                 167/udp   NAMP
#                    Marty Schoffstahl <schoff@NISC.NYSER.NET>
rsvd                 168/tcp   RSVD
rsvd                 168/udp   RSVD
#                    Neil Todd <mcvax!ist.co.uk!neil@UUNET.UU.NET>
send                 169/tcp   SEND
send                 169/udp   SEND
#                    William D. Wisner <wisner@HAYES.FAI.ALASKA.EDU>
print-srv            170/tcp   Network PostScript
print-srv            170/udp   Network PostScript
#                    Brian Reid <reid@DECWRL.DEC.COM>
multiplex            171/tcp   Network Innovations Multiplex
multiplex            171/udp   Network Innovations Multiplex
cl/1                 172/tcp   Network Innovations CL/1
cl/1                 172/udp   Network Innovations CL/1
#                    Kevin DeVault <<---none--->
xyplex-mux           173/tcp   Xyplex
xyplex-mux           173/udp   Xyplex
#                    Bob Stewart <STEWART@XYPLEX.COM>
mailq                174/tcp   MAILQ
mailq                174/udp   MAILQ
#                    Rayan Zachariassen <rayan@AI.TORONTO.EDU>
vmnet                175/tcp   VMNET
vmnet                175/udp   VMNET
#                    Christopher Tengi <tengi@Princeton.EDU>
```

```
genrad-mux              176/tcp    GENRAD-MUX
genrad-mux              176/udp    GENRAD-MUX
#             Ron Thornton <thornton@qm7501.genrad.com>
xdmcp                   177/tcp    X Display Manager Control Protocol
xdmcp                   177/udp    X Display Manager Control Protocol
#             Robert W. Scheifler <RWS@XX.LCS.MIT.EDU>
nextstep                178/tcp    NextStep Window Server
nextstep                178/udp    NextStep Window Server
#             Leo Hourvitz <leo@NEXT.COM>
bgp                     179/tcp    Border Gateway Protocol
bgp                     179/udp    Border Gateway Protocol
#             Kirk Lougheed <LOUGHEED@MATHOM.CISCO.COM>
ris                     180/tcp    Intergraph
ris                     180/udp    Intergraph
#             Dave Buehmann <ingr!daveb@UUNET.UU.NET>
unify                   181/tcp    Unify
unify                   181/udp    Unify
#             Mark Ainsley <ianaportmaster@unify.com>
audit                   182/tcp    Unisys Audit SITP
audit                   182/udp    Unisys Audit SITP
#             Gil Greenbaum <gcole@nisd.cam.unisys.com>
ocbinder                183/tcp    OCBinder
ocbinder                183/udp    OCBinder
ocserver                184/tcp    OCServer
ocserver                184/udp    OCServer
#             Jerrilynn Okamura <--none--->
remote-kis              185/tcp    Remote-KIS
remote-kis              185/udp    Remote-KIS
kis                     186/tcp    KIS Protocol
kis                     186/udp    KIS Protocol
#             Ralph Droms <rdroms@NRI.RESTON.VA.US>
aci                     187/tcp    Application Communication Interface
aci                     187/udp    Application Communication Interface
#             Rick Carlos <rick.ticipa.csc.ti.com>
mumps                   188/tcp    Plus Five's MUMPS
mumps                   188/udp    Plus Five's MUMPS
#             Hokey Stenn <hokey@PLUS5.COM>
qft                     189/tcp    Queued File Transport
qft                     189/udp    Queued File Transport
#             Wayne Schroeder <schroeder@SDS.SDSC.EDU>
gacp                    190/tcp    Gateway Access Control Protocol
gacp                    190/udp    Gateway Access Control Protocol
#             C. Philip Wood <cpw@LANL.GOV>
prospero                191/tcp    Prospero Directory Service
prospero                191/udp    Prospero Directory Service
#             B. Clifford Neuman <bcn@isi.edu>
```

osu-nms	192/tcp	OSU Network Monitoring System
osu-nms	192/udp	OSU Network Monitoring System
#	Doug Karl <KARL-D@OSU-20.IRCC.OHIO-STATE.EDU>	
srmp	193/tcp	Spider Remote Monitoring Protocol
srmp	193/udp	Spider Remote Monitoring Protocol
#	Ted J. Socolofsky <Teds@SPIDER.CO.UK>	
irc	194/tcp	Internet Relay Chat Protocol
irc	194/udp	Internet Relay Chat Protocol
#	Jarkko Oikarinen <jto@TOLSUN.OULU.FI>	
dn6-nlm-aud	195/tcp	DNSIX Network Level Module Audit
dn6-nlm-aud	195/udp	DNSIX Network Level Module Audit
dn6-smm-red	196/tcp	DNSIX Session Mgt Module Audit Redir
dn6-smm-red	196/udp	DNSIX Session Mgt Module Audit Redir
#	Lawrence Lebahn <DIA3@PAXRV-NES.NAVY.MIL>	
dls	197/tcp	Directory Location Service
dls	197/udp	Directory Location Service
dls-mon	198/tcp	Directory Location Service Monitor
dls-mon	198/udp	Directory Location Service Monitor
#	Scott Bellew <smb@cs.purdue.edu>	
smux	199/tcp	SMUX
smux	199/udp	SMUX
#	Marshall Rose <mrose@dbc.mtview.ca.us>	
src	200/tcp	IBM System Resource Controller
src	200/udp	IBM System Resource Controller
#	Gerald McBrearty <---none--->	
at-rtmp	201/tcp	AppleTalk Routing Maintenance
at-rtmp	201/udp	AppleTalk Routing Maintenance
at-nbp	202/tcp	AppleTalk Name Binding
at-nbp	202/udp	AppleTalk Name Binding
at-3	203/tcp	AppleTalk Unused
at-3	203/udp	AppleTalk Unused
at-echo	204/tcp	AppleTalk Echo
at-echo	204/udp	AppleTalk Echo
at-5	205/tcp	AppleTalk Unused
at-5	205/udp	AppleTalk Unused
at-zis	206/tcp	AppleTalk Zone Information
at-zis	206/udp	AppleTalk Zone Information
at-7	207/tcp	AppleTalk Unused
at-7	207/udp	AppleTalk Unused
at-8	208/tcp	AppleTalk Unused
at-8	208/udp	AppleTalk Unused
#	Rob Chandhok <chandhok@gnome.cs.cmu.edu>	
qmtp	209/tcp	The Quick Mail Transfer Protocol
qmtp	209/udp	The Quick Mail Transfer Protocol
#	Dan Bernstein <djb@silverton.berkeley.edu>	
z39.50	210/tcp	ANSI Z39.50

```
z39.50                      210/udp   ANSI Z39.50
#                  Mark H. Needleman <markn@sirsi.com>
914c/g                      211/tcp   Texas Instruments 914C/G Terminal
914c/g                      211/udp   Texas Instruments 914C/G Terminal
#                  Bill Harrell <---none--->
anet                        212/tcp   ATEXSSTR
anet                        212/udp   ATEXSSTR
#                  Jim Taylor <taylor@heart.epps.kodak.com>
ipx                         213/tcp   IPX
ipx                         213/udp   IPX
#                  Don Provan <donp@xlnvax.novell.com>
vmpwscs                     214/tcp   VM PWSCS
vmpwscs                     214/udp   VM PWSCS
#                  Dan Shia <dset!shia@uunet.UU.NET>
softpc                      215/tcp   Insignia Solutions
softpc                      215/udp   Insignia Solutions
#                  Martyn Thomas <---none--->
CAllic                      216/tcp   Computer Associates Int'l License
Server
CAllic                      216/udp   Computer Associates Int'l License
Server
#                  Chuck Spitz <spich04@cai.com>
dbase                       217/tcp   dBASE Unix
dbase                       217/udp   dBASE Unix
#                  Don Gibson
#          <sequent!aero!twinsun!ashtate.A-T.COM!dong@uunet.UU.NET>
mpp                         218/tcp   Netix Message Posting Protocol
mpp                         218/udp   Netix Message Posting Protocol
#                  Shannon Yeh <yeh@netix.com>
uarps                       219/tcp   Unisys ARPs
uarps                       219/udp   Unisys ARPs
#                  Ashok Marwaha <---none--->
imap3                       220/tcp   Interactive Mail Access Protocol v3
imap3                       220/udp   Interactive Mail Access Protocol v3
#                  James Rice <RICE@SUMEX-AIM.STANFORD.EDU>
fln-spx                     221/tcp   Berkeley rlogind with SPX auth
fln-spx                     221/udp   Berkeley rlogind with SPX auth
rsh-spx                     222/tcp   Berkeley rshd with SPX auth
rsh-spx                     222/udp   Berkeley rshd with SPX auth
cdc                         223/tcp   Certificate Distribution Center
cdc                         223/udp   Certificate Distribution Center
#          Kannan Alagappan <kannan@sejour.enet.dec.com>
########### Possible Conflict of Port 222 with "Masqdialer"#############
### Contact for Masqdialer is Charles Wright <cpwright@villagenet.com>###
masqdialer                  224/tcp   masqdialer
masqdialer                  224/udp   masqdialer
```

```
#                      Charles Wright <cpwright@villagenet.com>
#                      225-241    Reserved
#                 Jon Postel <postel@isi.edu>
direct                 242/tcp    Direct
direct                 242/udp    Direct
#                 Herb Sutter <HerbS@cntc.com>
sur-meas               243/tcp    Survey Measurement
sur-meas               243/udp    Survey Measurement
#                 Dave Clark <ddc@LCS.MIT.EDU>
inbusiness             244/tcp    inbusiness
inbusiness             244/udp    inbusiness
#                 Derrick Hisatake <derrick.i.hisatake@intel.com>
link                   245/tcp    LINK
link                   245/udp    LINK
dsp3270                246/tcp    Display Systems Protocol
dsp3270                246/udp    Display Systems Protocol
#                 Weldon J. Showalter <Gamma@MINTAKA.DCA.MIL>
subntbcst_tftp         247/tcp    SUBNTBCST_TFTP
subntbcst_tftp         247/udp    SUBNTBCST_TFTP
#                      John Fake <fake@us.ibm.com>
bhfhs                  248/tcp    bhfhs
bhfhs                  248/udp    bhfhs
#                      John Kelly <johnk@bellhow.com>
#                      249-255    Reserved
#                 Jon Postel <postel@isi.edu>
rap                    256/tcp    RAP
rap                    256/udp    RAP
#                 J.S. Greenfield <greeny@raleigh.ibm.com>
set                    257/tcp    Secure Electronic Transaction
set                    257/udp    Secure Electronic Transaction
#                 Donald Eastlake <dee3@torque.pothole.com>
yak-chat               258/tcp    Yak Winsock Personal Chat
yak-chat               258/udp    Yak Winsock Personal Chat
#                 Brian Bandy <bbandy@swbell.net>
esro-gen               259/tcp    Efficient Short Remote Operations
esro-gen               259/udp    Efficient Short Remote Operations
#                 Mohsen Banan <mohsen@rostam.neda.com>
openport               260/tcp    Openport
openport               260/udp    Openport
#                 John Marland <jmarland@dean.openport.com>
nsiiops                261/tcp    IIOP Name Service over TLS/SSL
nsiiops                261/udp    IIOP Name Service over TLS/SSL
#                 Jeff Stewart <jstewart@netscape.com>
arcisdms               262/tcp    Arcisdms
arcisdms               262/udp    Arcisdms
#                      Russell Crook (rmc@sni.ca>
```

```
hdap                          263/tcp   HDAP
hdap                          263/udp   HDAP
#                    Troy Gau <troy@zyxel.com>
bgmp                          264/tcp   BGMP
bgmp                          264/udp   BGMP
#                    Dave Thaler <thalerd@eecs.umich.edu>
x-bone-ctl                    265/tcp   X-Bone CTL
x-bone-ctl                    265/udp   X-Bone CTL
#                    Joe Touch <touch@isi.edu>
sst                           266/tcp   SCSI on ST
sst                           266/udp   SCSI on ST
#                    Donald D. Woelz <don@genroco.com>
td-service                    267/tcp   Tobit David Service Layer
td-service                    267/udp   Tobit David Service Layer
td-replica                    268/tcp   Tobit David Replica
td-replica                    268/udp   Tobit David Replica
#                    Franz-Josef Leuders <development@tobit.com>
#                             269-279   Unassigned
http-mgmt                     280/tcp   http-mgmt
http-mgmt                     280/udp   http-mgmt
#                    Adrian Pell
#                    <PELL_ADRIAN/HP-UnitedKingdom_om6@hplb.hpl.hp.com>
personal-link                 281/tcp   Personal Link
personal-link                 281/udp   Personal Link
#                    Dan Cummings <doc@cnr.com>
cableport-ax                  282/tcp   Cable Port A/X
cableport-ax                  282/udp   Cable Port A/X
#                    Craig Langfahl <Craig_J_Langfahl@ccm.ch.intel.com>
rescap                        283/tcp   rescap
rescap                        283/udp   rescap
#                    Paul Hoffman <phoffman@imc.org>
corerjd                       284/tcp   corerjd
corerjd                       284/udp   corerjd
#                    Chris Thornhill <port_contact@cjt.ca>
#                             285       Unassigned
fxp                           286/tcp   FXP Communication
fxp                           286/udp   FXP Communication
#                    James Darnall <james_r_darnall@sbcglobal.net>
k-block                       287/tcp   K-BLOCK
k-block                       287/udp   K-BLOCK
#                    Simon P Jackson <jacko@kring.co.uk>
#                             288-307   Unassigned
novastorbakcup                308/tcp   Novastor Backup
novastorbakcup                308/udp   Novastor Backup
#                    Brian Dickman <brian@novastor.com>
entrusttime                   309/tcp   EntrustTime
```

```
entrusttime                    309/udp    EntrustTime
#                Peter Whittaker <pww@entrust.com>
bhmds                          310/tcp    bhmds
bhmds                          310/udp    bhmds
#                John Kelly <johnk@bellhow.com>
asip-webadmin                  311/tcp    AppleShare IP WebAdmin
asip-webadmin                  311/udp    AppleShare IP WebAdmin
#                Ann Huang <annhuang@apple.com>
vslmp                          312/tcp    VSLMP
vslmp                          312/udp    VSLMP
#                Gerben Wierda <Gerben_Wierda@RnA.nl>
magenta-logic                  313/tcp    Magenta Logic
magenta-logic                  313/udp    Magenta Logic
#                Karl Rousseau <kr@netfusion.co.uk>
opalis-robot                   314/tcp    Opalis Robot
opalis-robot                   314/udp    Opalis Robot
#                Laurent Domenech, Opalis
<ldomenech@opalis.com>
dpsi                           315/tcp    DPSI
dpsi                           315/udp    DPSI
#                Tony Scamurra <Tony@DesktopPaging.com>
decauth                        316/tcp    decAuth
decauth                        316/udp    decAuth
#                Michael Agishtein <misha@unx.dec.com>
zannet                         317/tcp    Zannet
zannet                         317/udp    Zannet
#                Zan Oliphant <zan@accessone.com>
pkix-timestamp                 318/tcp    PKIX TimeStamp
pkix-timestamp                 318/udp    PKIX TimeStamp
#                Robert Zuccherato
<robert.zuccherato@entrust.com>
ptp-event                      319/tcp    PTP Event
ptp-event                      319/udp    PTP Event
ptp-general                    320/tcp    PTP General
ptp-general                    320/udp    PTP General
#                John Eidson <eidson@hpl.hp.com>
pip                            321/tcp    PIP
pip                            321/udp    PIP
#                Gordon Mohr <gojomo@usa.net>
rtsps                          322/tcp    RTSPS
rtsps                          322/udp    RTSPS
#                Anders Klemets <anderskl@microsoft.com>
#                              323-332    Unassigned
texar                          333/tcp    Texar Security Port
texar                          333/udp    Texar Security Port
#                Eugen Bacic <ebacic@texar.com>
```

```
#                               334-343   Unassigned
pdap                            344/tcp   Prospero Data Access Protocol
pdap                            344/udp   Prospero Data Access Protocol
#               B. Clifford Neuman <bcn@isi.edu>
pawserv                         345/tcp   Perf Analysis Workbench
pawserv                         345/udp   Perf Analysis Workbench
zserv                           346/tcp   Zebra server
zserv                           346/udp   Zebra server
fatserv                         347/tcp   Fatmen Server
fatserv                         347/udp   Fatmen Server
csi-sgwp                        348/tcp   Cabletron Management Protocol
csi-sgwp                        348/udp   Cabletron Management Protocol
mftp                            349/tcp   mftp
mftp                            349/udp   mftp
#               Dave Feinleib <davefe@microsoft.com>
matip-type-a                    350/tcp   MATIP Type A
matip-type-a                    350/udp   MATIP Type A
matip-type-b                    351/tcp   MATIP Type B
matip-type-b                    351/udp   MATIP Type B
#               Alain Robert <arobert@par.sita.int>
# The following entry records an unassigned but widespread use
bhoetty                         351/tcp         bhoetty (added 5/21/97)
bhoetty                         351/udp   bhoetty
#               John Kelly <johnk@bellhow.com>
dtag-ste-sb                     352/tcp         DTAG (assigned long ago)
dtag-ste-sb                     352/udp         DTAG
#               Ruediger Wald <wald@ez-darmstadt.telekom.de>
# The following entry records an unassigned but widespread use
bhoedap4                        352/tcp   bhoedap4 (added 5/21/97)
bhoedap4                        352/udp   bhoedap4
#               John Kelly <johnk@bellhow.com>
ndsauth                         353/tcp   NDSAUTH
ndsauth                         353/udp   NDSAUTH
#               Jayakumar Ramalingam <jayakumar@novell.com>
bh611                           354/tcp   bh611
bh611                           354/udp   bh611
#               John Kelly <johnk@bellhow.com>
datex-asn                       355/tcp   DATEX-ASN
datex-asn                       355/udp   DATEX-ASN
#               Kenneth Vaughn <kvaughn@mail.viggen.com>
cloanto-net-1                   356/tcp   Cloanto Net 1
cloanto-net-1                   356/udp   Cloanto Net 1
#               Michael Battilana <mcb-iana@cloanto.com>
bhevent                         357/tcp   bhevent
bhevent                         357/udp   bhevent
#               John Kelly <johnk@bellhow.com>
```

shrinkwrap	358/tcp	Shrinkwrap
shrinkwrap	358/udp	Shrinkwrap
#	Bill Simpson <wsimpson@greendragon.com>	
nsrmp	359/tcp	Network Security Risk Management Protocol
nsrmp	359/udp	Network Security Risk Management Protocol
#	Eric Jacksch <jacksch@tenebris.ca>	
scoi2odialog	360/tcp	scoi2odialog
scoi2odialog	360/udp	scoi2odialog
#	Keith Petley <keithp@sco.COM>	
semantix	361/tcp	Semantix
semantix	361/udp	Semantix
#	Semantix <xsSupport@semantix.com>	
srssend	362/tcp	SRS Send
srssend	362/udp	SRS Send
#	Curt Mayer <curt@emergent.com>	
rsvp_tunnel	363/tcp	RSVP Tunnel
rsvp_tunnel	363/udp	RSVP Tunnel
#	Andreas Terzis <terzis@cs.ucla.edu>	
aurora-cmgr	364/tcp	Aurora CMGR
aurora-cmgr	364/udp	Aurora CMGR
#	Philip Budne <budne@auroratech.com>	
dtk	365/tcp	DTK
dtk	365/udp	DTK
#	Fred Cohen <fc@all.net>	
odmr	366/tcp	ODMR
odmr	366/udp	ODMR
#	Randall Gellens <randy@qualcomm.com>	
mortgageware	367/tcp	MortgageWare
mortgageware	367/udp	MortgageWare
#	Ole Hellevik <oleh@interlinq.com>	
qbikgdp	368/tcp	QbikGDP
qbikgdp	368/udp	QbikGDP
#	Adrien de Croy <adrien@qbik.com>	
rpc2portmap	369/tcp	rpc2portmap
rpc2portmap	369/udp	rpc2portmap
codaauth2	370/tcp	codaauth2
codaauth2	370/udp	codaauth2
#	Robert Watson <robert@cyrus.watson.org>	
clearcase	371/tcp	Clearcase
clearcase	371/udp	Clearcase
#	Dave LeBlang <leglang@atria.com>	
ulistproc	372/tcp	ListProcessor
ulistproc	372/udp	ListProcessor
#	Anastasios Kotsikonas <tasos@cs.bu.edu>	

legent-1	373/tcp	Legent Corporation
legent-1	373/udp	Legent Corporation
legent-2	374/tcp	Legent Corporation
legent-2	374/udp	Legent Corporation
#	Keith Boyce <---none--->	
hassle	375/tcp	Hassle
hassle	375/udp	Hassle
#	Reinhard Doelz <doelz@comp.bioz.unibas.ch>	
nip	376/tcp	Amiga Envoy Network Inquiry Proto
nip	376/udp	Amiga Envoy Network Inquiry Proto
#	Heinz Wrobel <hwrobel@gmx.de>	
tnETOS	377/tcp	NEC Corporation
tnETOS	377/udp	NEC Corporation
dsETOS	378/tcp	NEC Corporation
dsETOS	378/udp	NEC Corporation
#	Tomoo Fujita <tf@ar.bs1.fc.nec.co.jp>	
is99c	379/tcp	TIA/EIA/IS-99 modem client
is99c	379/udp	TIA/EIA/IS-99 modem client
is99s	380/tcp	TIA/EIA/IS-99 modem server
is99s	380/udp	TIA/EIA/IS-99 modem server
#	Frank Quick <fquick@qualcomm.com>	
hp-collector	381/tcp	hp performance data collector
hp-collector	381/udp	hp performance data collector
hp-managed-node	382/tcp	hp performance data managed node
hp-managed-node	382/udp	hp performance data managed node
hp-alarm-mgr	383/tcp	hp performance data alarm manager
hp-alarm-mgr	383/udp	hp performance data alarm manager
#	Frank Blakely <frankb@hpptc16.rose.hp.com>	
arns	384/tcp	A Remote Network Server System
arns	384/udp	A Remote Network Server System
#	David Hornsby <djh@munnari.OZ.AU>	
ibm-app	385/tcp	IBM Application
ibm-app	385/udp	IBM Application
#	Lisa Tomita <---none--->	
asa	386/tcp	ASA Message Router Object Def.
asa	386/udp	ASA Message Router Object Def.

#	Steve Laitinen <laitinen@brutus.aa.ab.com>		
aurp		387/tcp	Appletalk Update-Based Routing Pro.
aurp		387/udp	Appletalk Update-Based Routing Pro.
#	Chris Ranch <cranch@novell.com>		
unidata-ldm		388/tcp	Unidata LDM
unidata-ldm		388/udp	Unidata LDM
#	Glenn Davis <support@unidata.ucar.edu>		
ldap		389/tcp	Lightweight Directory Access Protocol
ldap		389/udp	Lightweight Directory Access Protocol
#	Tim Howes <Tim.Howes@terminator.cc.umich.edu>		
uis		390/tcp	UIS
uis		390/udp	UIS
#	Ed Barron <---none--->		
synotics-relay		391/tcp	SynOptics SNMP Relay Port
synotics-relay		391/udp	SynOptics SNMP Relay Port
synotics-broker		392/tcp	SynOptics Port Broker Port
synotics-broker		392/udp	SynOptics Port Broker Port
#	Illan Raab <iraab@synoptics.com>		
meta5		393/tcp	Meta5
meta5		393/udp	Meta5
#	Jim Kanzler <jim.kanzler@meta5.com>		
embl-ndt		394/tcp	EMBL Nucleic Data Transfer
embl-ndt		394/udp	EMBL Nucleic Data Transfer
#	Peter Gad <peter@bmc.uu.se>		
netcp		395/tcp	NETscout Control Protocol
netcp		395/udp	NETscout Control Protocol
#	Anil Singhal <---none--->		
netware-ip		396/tcp	Novell Netware over IP
netware-ip		396/udp	Novell Netware over IP
mptn		397/tcp	Multi Protocol Trans. Net.
mptn		397/udp	Multi Protocol Trans. Net.
#	Soumitra Sarkar <sarkar@vnet.ibm.com>		
kryptolan		398/tcp	Kryptolan
kryptolan		398/udp	Kryptolan
#	Peter de Laval <pdl@sectra.se>		
iso-tsap-c2		399/tcp	ISO Transport Class 2 Non-Control over TCP
iso-tsap-c2		399/udp	ISO Transport Class 2 Non-Control over UDP
#	Yanick Pouffary <pouffary@taec.enet.dec.com>		
work-sol		400/tcp	Workstation Solutions
work-sol		400/udp	Workstation Solutions

#	Jim Ward <jimw@worksta.com>		
ups		401/tcp	Uninterruptible Power Supply
ups		401/udp	Uninterruptible Power Supply
#	Charles Bennett <chuck@benatong.com>		
genie		402/tcp	Genie Protocol
genie		402/udp	Genie Protocol
#	Mark Hankin <---none--->		
decap		403/tcp	decap
decap		403/udp	decap
nced		404/tcp	nced
nced		404/udp	nced
ncld		405/tcp	ncld
ncld		405/udp	ncld
#	Richard Jones <---none--->		
imsp		406/tcp	Interactive Mail Support Protocol
imsp		406/udp	Interactive Mail Support Protocol
#	John Myers <jgm+@cmu.edu>		
timbuktu		407/tcp	Timbuktu
timbuktu		407/udp	Timbuktu
#	Marc Epard <marc@netopia.com>		
prm-sm		408/tcp	Prospero Resource Manager Sys. Man.
prm-sm		408/udp	Prospero Resource Manager Sys. Man.
prm-nm		409/tcp	Prospero Resource Manager Node Man.
prm-nm		409/udp	Prospero Resource Manager Node Man.
#	B. Clifford Neuman <bcn@isi.edu>		
decladebug		410/tcp	DECLadebug Remote Debug Protocol
decladebug		410/udp	DECLadebug Remote Debug Protocol
#	Anthony Berent <anthony.berent@reo.mts.dec.com>		
rmt		411/tcp	Remote MT Protocol
rmt		411/udp	Remote MT Protocol
#	Peter Eriksson <pen@lysator.liu.se>		
synoptics-trap		412/tcp	Trap Convention Port
synoptics-trap		412/udp	Trap Convention Port
#	Illan Raab <iraab@synoptics.com>		
smsp		413/tcp	Storage Management Services Protocol
smsp		413/udp	Storage Management Services Protocol
#	Murthy Srinivas <murthy@novell.com>		
infoseek		414/tcp	InfoSeek

infoseek	414/udp	InfoSeek
#	Steve Kirsch <stk@infoseek.com>	
bnet	415/tcp	BNet
bnet	415/udp	BNet
#	Jim Mertz <JMertz+RV09@rvdc.unisys.com>	
silverplatter	416/tcp	Silverplatter
silverplatter	416/udp	Silverplatter
#	Peter Ciuffetti <petec@silverplatter.com>	
onmux	417/tcp	Onmux
onmux	417/udp	Onmux
#	Stephen Hanna <hanna@world.std.com>	
hyper-g	418/tcp	Hyper-G
hyper-g	418/udp	Hyper-G
#	Frank Kappe <fkappe@iicm.tu-graz.ac.at>	
ariel1	419/tcp	Ariel 1
ariel1	419/udp	Ariel 1
#	Joel Karafin <jkarafin@infotrieve.com>	
smpte	420/tcp	SMPTE
smpte	420/udp	SMPTE
#	Si Becker <71362.22@CompuServe.COM>	
ariel2	421/tcp	Ariel 2
ariel2	421/udp	Ariel 2
ariel3	422/tcp	Ariel 3
ariel3	422/udp	Ariel 3
#	Joel Karafin <jkarafin@infotrieve.com>	
opc-job-start	423/tcp	IBM Operations Planning and Control Start
opc-job-start	423/udp	IBM Operations Planning and Control Start
opc-job-track	424/tcp	IBM Operations Planning and Control Track
opc-job-track	424/udp	IBM Operations Planning and Control Track
#	Conny Larsson <cocke@VNET.IBM.COM>	
icad-el	425/tcp	ICAD
icad-el	425/udp	ICAD
#	Larry Stone <lcs@icad.com>	
smartsdp	426/tcp	smartsdp
smartsdp	426/udp	smartsdp
#	Alexander Dupuy <dupuy@smarts.com>	
svrloc	427/tcp	Server Location
svrloc	427/udp	Server Location
#	<veizades@ftp.com>	
ocs_cmu	428/tcp	OCS_CMU
ocs_cmu	428/udp	OCS_CMU
ocs_amu	429/tcp	OCS_AMU

Modern Communications Systems

ocs_amu	429/udp	OCS_AMU
#	Florence Wyman <wyman@peabody.plk.af.mil>	
utmpsd	430/tcp	UTMPSD
utmpsd	430/udp	UTMPSD
utmpcd	431/tcp	UTMPCD
utmpcd	431/udp	UTMPCD
iasd	432/tcp	IASD
iasd	432/udp	IASD
#	Nir Baroz <nbaroz@encore.com>	
nnsp	433/tcp	NNSP
nnsp	433/udp	NNSP
#	Rob Robertson <rob@gangrene.berkeley.edu>	
mobileip-agent	434/tcp	MobileIP-Agent
mobileip-agent	434/udp	MobileIP-Agent
mobilip-mn	435/tcp	MobilIP-MN
mobilip-mn	435/udp	MobilIP-MN
#	Kannan Alagappan <kannan@sejour.lkg.dec.com>	
dna-cml	436/tcp	DNA-CML
dna-cml	436/udp	DNA-CML
#	Dan Flowers <flowers@smaug.lkg.dec.com>	
comscm	437/tcp	comscm
comscm	437/udp	comscm
#	Jim Teague <teague@zso.dec.com>	
dsfgw	438/tcp	dsfgw
dsfgw	438/udp	dsfgw
#	Andy McKeen <mckeen@osf.org>	
dasp	439/tcp	dasp Thomas Obermair
dasp	439/udp	dasp tommy@inlab.m.eunet.de
#	Thomas Obermair <tommy@inlab.m.eunet.de>	
sgcp	440/tcp	sgcp
sgcp	440/udp	sgcp
#	Marshall Rose <mrose@dbc.mtview.ca.us>	
decvms-sysmgt	441/tcp	decvms-sysmgt
decvms-sysmgt	441/udp	decvms-sysmgt
#	Lee Barton <barton@star.enet.dec.com>	
cvc_hostd	442/tcp	cvc_hostd
cvc_hostd	442/udp	cvc_hostd
#	Bill Davidson <billd@equalizer.cray.com>	
https	443/tcp	http protocol over TLS/SSL
https	443/udp	http protocol over TLS/SSL
#	Kipp E.B. Hickman <kipp@mcom.com>	
snpp	444/tcp	Simple Network Paging Protocol
snpp	444/udp	Simple Network Paging Protocol
#	[RFC1568]	
microsoft-ds	445/tcp	Microsoft-DS
microsoft-ds	445/udp	Microsoft-DS

#		Pradeep Bahl <pradeepb@microsoft.com>	
ddm-rdb	446/tcp	DDM-Remote Relational Database Access	
ddm-rdb	446/udp	DDM-Remote Relational Database Access	
ddm-dfm	447/tcp	DDM-Distributed File Management	
ddm-dfm	447/udp	DDM-Distributed File Management	
#		Steven Ritland <srr@us.ibm.com>	
ddm-ssl	448/tcp	DDM-Remote DB Access Using Secure Sockets	
ddm-ssl	448/udp	DDM-Remote DB Access Using Secure Sockets	
#		Steven Ritland <srr@us.ibm.com>	
as-servermap	449/tcp	AS Server Mapper	
as-servermap	449/udp	AS Server Mapper	
#		Barbara Foss <BGFOSS@rchvmv.vnet.ibm.com>	
tserver	450/tcp	Computer Supported Telecomunication Applications	
tserver	450/udp	Computer Supported Telecomunication Applications	
#		Harvey S. Schultz <harvey@acm.org>	
sfs-smp-net	451/tcp	Cray Network Semaphore server	
sfs-smp-net	451/udp	Cray Network Semaphore server	
sfs-config	452/tcp	Cray SFS config server	
sfs-config	452/udp	Cray SFS config server	
#		Walter Poxon <wdp@ironwood.cray.com>	
creativeserver	453/tcp	CreativeServer	
creativeserver	453/udp	CreativeServer	
contentserver	454/tcp	ContentServer	
contentserver	454/udp	ContentServer	
creativepartnr	455/tcp	CreativePartnr	
creativepartnr	455/udp	CreativePartnr	
#		Jesus Ortiz <jesus_ortiz@emotion.com>	
macon-tcp	456/tcp	macon-tcp	
macon-udp	456/udp	macon-udp	
#		Yoshinobu Inoue	
#		<shin@hodaka.mfd.cs.fujitsu.co.jp>	
scohelp	457/tcp	scohelp	
scohelp	457/udp	scohelp	
#		Faith Zack <faithz@sco.com>	
appleqtc	458/tcp	apple quick time	
appleqtc	458/udp	apple quick time	
#		Murali Ranganathan	
#		<murali_ranganathan@@quickmail.apple.com>	
ampr-rcmd	459/tcp	ampr-rcmd	
ampr-rcmd	459/udp	ampr-rcmd	

```
#                Rob Janssen <rob@sys3.pe1chl.ampr.org>
skronk                        460/tcp    skronk
skronk                        460/udp    skronk
#                Henry Strickland <strick@yak.net>
datasurfsrv                   461/tcp    DataRampSrv
datasurfsrv                   461/udp    DataRampSrv
datasurfsrvsec                462/tcp    DataRampSrvSec
datasurfsrvsec                462/udp    DataRampSrvSec
#                Diane Downie <downie@jibe.MV.COM>
alpes                         463/tcp    alpes
alpes                         463/udp    alpes
#                Alain Durand <Alain.Durand@imag.fr>
kpasswd                       464/tcp    kpasswd
kpasswd                       464/udp    kpasswd
#                Theodore Ts'o <tytso@MIT.EDU>
urd                           465/tcp    URL Rendesvous Directory for SSM
igmpv3lite                    465/udp    IGMP over UDP for SSM
#                Toerless Eckert <eckert@cisco.com>
digital-vrc                   466/tcp    digital-vrc
digital-vrc                   466/udp    digital-vrc
#                Peter Higginson <higginson@mail.dec.com>
mylex-mapd                    467/tcp    mylex-mapd
mylex-mapd                    467/udp    mylex-mapd
#                Gary Lewis <GaryL@hq.mylex.com>
photuris                      468/tcp    proturis
photuris                      468/udp    proturis
#                Bill Simpson <Bill.Simpson@um.cc.umich.edu>
rcp                           469/tcp    Radio Control Protocol
rcp                           469/udp    Radio Control Protocol
#                Jim Jennings +1-708-538-7241
scx-proxy                     470/tcp    scx-proxy
scx-proxy                     470/udp    scx-proxy
#                Scott Narveson <sjn@cray.com>
mondex                        471/tcp    Mondex
mondex                        471/udp    Mondex
#                Bill Reding <redingb@nwdt.natwest.co.uk>
ljk-login                     472/tcp    ljk-login
ljk-login                     472/udp    ljk-login
#                LJK Software, Cambridge, Massachusetts
#                <support@ljk.com>
hybrid-pop                    473/tcp    hybrid-pop
hybrid-pop                    473/udp    hybrid-pop
#                Rami Rubin <rami@hybrid.com>
tn-tl-w1                      474/tcp    tn-tl-w1
tn-tl-w2                      474/udp    tn-tl-w2
#                Ed Kress <eskress@thinknet.com>
```

tcpnethaspsrv	475/tcp	tcpnethaspsrv
tcpnethaspsrv	475/udp	tcpnethaspsrv
#	Charlie Hava <charlie@aladdin.co.il>	
tn-tl-fd1	476/tcp	tn-tl-fd1
tn-tl-fd1	476/udp	tn-tl-fd1
#	Ed Kress <eskress@thinknet.com>	
ss7ns	477/tcp	ss7ns
ss7ns	477/udp	ss7ns
#	Jean-Michel URSCH <ursch@taec.enet.dec.com>	
spsc	478/tcp	spsc
spsc	478/udp	spsc
#	Mike Rieker <mikea@sp32.com>	
iafserver	479/tcp	iafserver
iafserver	479/udp	iafserver
iafdbase	480/tcp	iafdbase
iafdbase	480/udp	iafdbase
#	ricky@solect.com <Rick Yazwinski>	
ph	481/tcp	Ph service
ph	481/udp	Ph service
#	Roland Hedberg <Roland.Hedberg@umdac.umu.se>	
bgs-nsi	482/tcp	bgs-nsi
bgs-nsi	482/udp	bgs-nsi
#	Jon Saperia <saperia@bgs.com>	
ulpnet	483/tcp	ulpnet
ulpnet	483/udp	ulpnet
#	Kevin Mooney <kevinm@bfs.unibol.com>	
integra-sme	484/tcp	Integra Software Management Environment
integra-sme	484/udp	Integra Software Management Environment
#	Randall Dow <rand@randix.m.isr.de>	
powerburst	485/tcp	Air Soft Power Burst
powerburst	485/udp	Air Soft Power Burst
#	<gary@airsoft.com>	
avian	486/tcp	avian
avian	486/udp	avian
#	Robert Ullmann	
#	<Robert_Ullmann/CAM/Lotus.LOTUS@crd.lotus.com>	
saft	487/tcp	saft Simple Asynchronous File Transfer
saft	487/udp	saft Simple Asynchronous File Transfer
#	Ulli Horlacher <framstag@rus.uni-stuttgart.de>	
gss-http	488/tcp	gss-http
gss-http	488/udp	gss-http
#	Doug Rosenthal <rosenthl@krypton.einet.net>	

nest-protocol	489/tcp	nest-protocol
nest-protocol	489/udp	nest-protocol
#	Gilles Gameiro <ggameiro@birdland.com>	
micom-pfs	490/tcp	micom-pfs
micom-pfs	490/udp	micom-pfs
#	David Misunas <DMisunas@micom.com>	
go-login	491/tcp	go-login
go-login	491/udp	go-login
#	Troy Morrison <troy@graphon.com>	
ticf-1	492/tcp	Transport Independent Convergence for FNA
ticf-1	492/udp	Transport Independent Convergence for FNA
ticf-2	493/tcp	Transport Independent Convergence for FNA
ticf-2	493/udp	Transport Independent Convergence for FNA
#	Mamoru Ito <Ito@pcnet.ks.pfu.co.jp>	
pov-ray	494/tcp	POV-Ray
pov-ray	494/udp	POV-Ray
#	POV-Team Co-ordinator	
#	<iana-port.remove-spamguard@povray.org>	
intecourier	495/tcp	intecourier
intecourier	495/udp	intecourier
#	Steve Favor <sfavor@tigger.intecom.com>	
pim-rp-disc	496/tcp	PIM-RP-DISC
pim-rp-disc	496/udp	PIM-RP-DISC
#	Dino Farinacci <dino@cisco.com>	
dantz	497/tcp	dantz
dantz	497/udp	dantz
#	Richard Zulch <richard_zulch@dantz.com>	
siam	498/tcp	siam
siam	498/udp	siam
#	Philippe Gilbert <pgilbert@cal.fr>	
iso-ill	499/tcp	ISO ILL Protocol
iso-ill	499/udp	ISO ILL Protocol
#	Mark H. Needleman <markn@sirsi.com>	
isakmp	500/tcp	isakmp
isakmp	500/udp	isakmp
#	Mark Schertler <mjs@tycho.ncsc.mil>	
stmf	501/tcp	STMF
stmf	501/udp	STMF
#	Alan Ungar <aungar@farradyne.com>	
asa-appl-proto	502/tcp	asa-appl-proto
asa-appl-proto	502/udp	asa-appl-proto
#	Dennis Dube <ddube@modicon.com>	

```
intrinsa                        503/tcp    Intrinsa
intrinsa                        503/udp    Intrinsa
#              Robert Ford <robert@intrinsa.com>
citadel                         504/tcp    citadel
citadel                         504/udp    citadel
#              Art Cancro <ajc@uncnsrd.mt-kisco.ny.us>
mailbox-lm                      505/tcp    mailbox-lm
mailbox-lm                      505/udp    mailbox-lm
#              Beverly Moody <Beverly_Moody@stercomm.com>
ohimsrv                         506/tcp    ohimsrv
ohimsrv                         506/udp    ohimsrv
#              Scott Powell <spowell@openhorizon.com>
crs                             507/tcp    crs
crs                             507/udp    crs
#              Brad Wright <bradwr@microsoft.com>
xvttp                           508/tcp    xvttp
xvttp                           508/udp    xvttp
#              Keith J. Alphonso <alphonso@ncs-ssc.com>
snare                           509/tcp    snare
snare                           509/udp    snare
#              Dennis Batchelder <dennis@capres.com>
fcp                             510/tcp    FirstClass Protocol
fcp                             510/udp    FirstClass Protocol
#              Mike Marshburn <paul@softarc.com>
passgo                          511/tcp    PassGo
passgo                          511/udp    PassGo
#              John Rainford <jrainford@passgo.com>
exec                            512/tcp    remote process execution;
#              authentication performed using
#              passwords and UNIX login names
comsat                          512/udp
biff                            512/udp    used by mail system to notify
                                              users
#              of new mail received; currently
#              receives messages only from
#              processes on the same machine
login                           513/tcp    remote login a la telnet;
#              automatic authentication performed
#              based on priviledged port numbers
#              and distributed data bases which
#              identify "authentication domains"
who                             513/udp    maintains data bases showing
                                              who's
#              logged in to machines on a local
#              net and the load average of the
#              machine
```

```
shell                            514/tcp   cmd
#             like exec, but automatic authentication
#             is performed as for login server
syslog                           514/udp
printer                          515/tcp   spooler
printer                          515/udp   spooler
videotex                         516/tcp   videotex
videotex                         516/udp   videotex
#             Daniel Mavrakis <system@venus.mctel.fr>
talk                             517/tcp   like tenex link, but across
#             machine - unfortunately, doesn't
#             use link protocol (this is actually
#             just a rendezvous port from which a
#             tcp connection is established)
talk                             517/udp   like tenex link, but across
#             machine - unfortunately, doesn't
#             use link protocol (this is actually
#             just a rendezvous port from which a
#             tcp connection is established)
ntalk                            518/tcp
ntalk                            518/udp
utime                            519/tcp   unixtime
utime                            519/udp   unixtime
efs                              520/tcp   extended file name server
router                           520/udp   local routing process (on site);
#             uses variant of Xerox NS routing
#             information protocol - RIP
ripng                            521/tcp   ripng
ripng                            521/udp   ripng
#             Robert E. Minnear <minnear@ipsilon.com>
ulp                              522/tcp   ULP
ulp                              522/udp   ULP
#             Max Morris <maxm@MICROSOFT.com>
ibm-db2                          523/tcp   IBM-DB2
ibm-db2                          523/udp   IBM-DB2
#             Juliana Hsu <jhsu@ca.ibm.com>
ncp                              524/tcp   NCP
ncp                              524/udp   NCP
#             Don Provan <donp@sjf.novell.com>
timed                            525/tcp   timeserver
timed                            525/udp   timeserver
tempo                            526/tcp   newdate
tempo                            526/udp   newdate
#             Unknown
stx                              527/tcp   Stock IXChange
stx                              527/udp   Stock IXChange
```

custix	528/tcp	Customer IXChange
custix	528/udp	Customer IXChange
#	Ferdi Ladeira <ferdi.ladeira@ixchange.com>	
irc-serv	529/tcp	IRC-SERV
irc-serv	529/udp	IRC-SERV
#	Brian Tackett <cym@acrux.net>	
courier	530/tcp	rpc
courier	530/udp	rpc
conference	531/tcp	chat
conference	531/udp	chat
netnews	532/tcp	readnews
netnews	532/udp	readnews
netwall	533/tcp	for emergency broadcasts
netwall	533/udp	for emergency broadcasts
mm-admin	534/tcp	MegaMedia Admin
mm-admin	534/udp	MegaMedia Admin
#	Andreas Heidemann <a.heidemann@ais-gmbh.de>	
iiop	535/tcp	iiop
iiop	535/udp	iiop
#	Jeff M.Michaud <michaud@zk3.dec.com>	
opalis-rdv	536/tcp	opalis-rdv
opalis-rdv	536/udp	opalis-rdv
#	Laurent Domenech <ldomenech@opalis.com>	
nmsp	537/tcp	Networked Media Streaming Protocol
nmsp	537/udp	Networked Media Streaming Protocol
#	Paul Santinelli Jr. <psantinelli@narrative.com>	
gdomap	538/tcp	gdomap
gdomap	538/udp	gdomap
#	Richard Frith-Macdonald <richard@brainstorm.co.uk>	
apertus-ldp	539/tcp	Apertus Technologies Load Determination
apertus-ldp	539/udp	Apertus Technologies Load Determination
uucp	540/tcp	uucpd
uucp	540/udp	uucpd
uucp-rlogin	541/tcp	uucp-rlogin
uucp-rlogin	541/udp	uucp-rlogin
#	Stuart Lynne <sl@wimsey.com>	
commerce	542/tcp	commerce
commerce	542/udp	commerce
#	Randy Epstein <repstein@host.net>	
klogin	543/tcp	
klogin	543/udp	
kshell	544/tcp	krcmd

kshell	544/udp	krcmd
appleqtcsrvr	545/tcp	appleqtcsrvr
appleqtcsrvr	545/udp	appleqtcsrvr
#	Murali Ranganathan	
#	<Murali_Ranganathan@quickmail.apple.com>	
dhcpv6-client	546/tcp	DHCPv6 Client
dhcpv6-client	546/udp	DHCPv6 Client
dhcpv6-server	547/tcp	DHCPv6 Server
dhcpv6-server	547/udp	DHCPv6 Server
#	Jim Bound <bound@zk3.dec.com>	
afpovertcp	548/tcp	AFP over TCP
afpovertcp	548/udp	AFP over TCP
#	Leland Wallace <randall@apple.com>	
idfp	549/tcp	IDFP
idfp	549/udp	IDFP
#	Ramana Kovi <ramana@kovi.com>	
new-rwho	550/tcp	new-who
new-rwho	550/udp	new-who
cybercash	551/tcp	cybercash
cybercash	551/udp	cybercash
#	Donald E. Eastlake 3rd <dee@cybercash.com>	
devshr-nts	552/tcp	DeviceShare
devshr-nts	552/udp	DeviceShare
#	Benjamin Rosenberg <brosenberg@advsyscon.com>	
pirp	553/tcp	pirp
pirp	553/udp	pirp
#	D. J. Bernstein <djb@silverton.berkeley.edu>	
rtsp	554/tcp	Real Time Stream Control Protocol
rtsp	554/udp	Real Time Stream Control Protocol
#	Rob Lanphier <robla@prognet.com>	
dsf	555/tcp	
dsf	555/udp	
remotefs	556/tcp	rfs server
remotefs	556/udp	rfs server
openvms-sysipc	557/tcp	openvms-sysipc
openvms-sysipc	557/udp	openvms-sysipc
#	Alan Potter <potter@movies.enet.dec.com>	
sdnskmp	558/tcp	SDNSKMP
sdnskmp	558/udp	SDNSKMP
teedtap	559/tcp	TEEDTAP
teedtap	559/udp	TEEDTAP
#	Mort Hoffman <hoffman@mail.ndhm.gtegsc.com>	
rmonitor	560/tcp	rmonitord
rmonitor	560/udp	rmonitord

monitor	561/tcp	
monitor	561/udp	
chshell	562/tcp	chcmd
chshell	562/udp	chcmd
nntps	563/tcp	nntp protocol over TLS/SSL (was snntp)
nntps	563/udp	nntp protocol over TLS/SSL (was snntp)
#	Kipp E.B. Hickman <kipp@netscape.com>	
9pfs	564/tcp	plan 9 file service
9pfs	564/udp	plan 9 file service
whoami	565/tcp	whoami
whoami	565/udp	whoami
streettalk	566/tcp	streettalk
streettalk	566/udp	streettalk
banyan-rpc	567/tcp	banyan-rpc
banyan-rpc	567/udp	banyan-rpc
#	Tom Lemaire <toml@banyan.com>	
ms-shuttle	568/tcp	microsoft shuttle
ms-shuttle	568/udp	microsoft shuttle
#	Rudolph Balaz <rudolphb@microsoft.com>	
ms-rome	569/tcp	microsoft rome
ms-rome	569/udp	microsoft rome
#	Rudolph Balaz <rudolphb@microsoft.com>	
meter	570/tcp	demon
meter	570/udp	demon
meter	571/tcp	udemon
meter	571/udp	udemon
sonar	572/tcp	sonar
sonar	572/udp	sonar
#	Keith Moore <moore@cs.utk.edu>	
banyan-vip	573/tcp	banyan-vip
banyan-vip	573/udp	banyan-vip
#	Denis Leclerc <DLeclerc@banyan.com>	
ftp-agent	574/tcp	FTP Software Agent System
ftp-agent	574/udp	FTP Software Agent System
#	Michael S. Greenberg <arnoff@ftp.com>	
vemmi	575/tcp	VEMMI
vemmi	575/udp	VEMMI
#	Daniel Mavrakis <mavrakis@mctel.fr>	
ipcd	576/tcp	ipcd
ipcd	576/udp	ipcd
vnas	577/tcp	vnas
vnas	577/udp	vnas
ipdd	578/tcp	ipdd
ipdd	578/udp	ipdd

#	Jay Farhat <jfarhat@ipass.com>	
decbsrv	579/tcp	decbsrv
decbsrv	579/udp	decbsrv
#	Rudi Martin	
<movies::martin"@movies.enet.dec.com>		
sntp-heartbeat	580/tcp	SNTP HEARTBEAT
sntp-heartbeat	580/udp	SNTP HEARTBEAT
#	Louis Mamakos <louie@uu.net>	
bdp	581/tcp	Bundle Discovery Protocol
bdp	581/udp	Bundle Discovery Protocol
#	Gary Malkin <gmalkin@xylogics.com>	
scc-security	582/tcp	SCC Security
scc-security	582/udp	SCC Security
#	Prashant Dholakia <prashant@semaphorecom.com>	
philips-vc	583/tcp	Philips Video-Conferencing
philips-vc	583/udp	Philips Video-Conferencing
#	Janna Chang <janna@pmc.philips.com>	
keyserver	584/tcp	Key Server
keyserver	584/udp	Key Server
#	Gary Howland <gary@systemics.com>	
imap4-ssl	585/tcp	IMAP4+SSL (use 993 instead)
imap4-ssl	585/udp	IMAP4+SSL (use 993 instead)
#	Terry Gray <gray@cac.washington.edu>	
#	Use of 585 is not recommended, use 993 instead	
password-chg	586/tcp	Password Change
password-chg	586/udp	Password Change
submission	587/tcp	Submission
submission	587/udp	Submission
#	Randy Gellens <randy@qualcomm.com>	
cal	588/tcp	CAL
cal	588/udp	CAL
#	Myron Hattig <Myron_Hattig@ccm.jf.intel.com>	
eyelink	589/tcp	EyeLink
eyelink	589/udp	EyeLink
#	Dave Stampe <dstampe@psych.toronto.edu>	
tns-cml	590/tcp	TNS CML
tns-cml	590/udp	TNS CML
#	Jerome Albin <albin@taec.enet.dec.com>	
http-alt	591/tcp	FileMaker, Inc. - HTTP Alternate (see Port 80)
http-alt	591/udp	FileMaker, Inc. - HTTP Alternate (see Port 80)
#	Clay Maeckel <clay_maeckel@filemaker.com>	
eudora-set	592/tcp	Eudora Set
eudora-set	592/udp	Eudora Set
#	Randall Gellens <randy@qualcomm.com>	

http-rpc-epmap	593/tcp	HTTP RPC Ep Map
http-rpc-epmap	593/udp	HTTP RPC Ep Map
#	Edward Reus <edwardr@microsoft.com>	
tpip	594/tcp	TPIP
tpip	594/udp	TPIP
#	Brad Spear <spear@platinum.com>	
cab-protocol	595/tcp	CAB Protocol
cab-protocol	595/udp	CAB Protocol
#	Winston Hetherington	
smsd	596/tcp	SMSD
smsd	596/udp	SMSD
#	Wayne Barlow <web@unx.dec.com>	
ptcnameservice	597/tcp	PTC Name Service
ptcnameservice	597/udp	PTC Name Service
#	Yuri Machkasov <yuri@ptc.com>	
sco-websrvrmg3	598/tcp	SCO Web Server Manager 3
sco-websrvrmg3	598/udp	SCO Web Server Manager 3
#	Simon Baldwin <simonb@sco.com>	
acp	599/tcp	Aeolon Core Protocol
acp	599/udp	Aeolon Core Protocol
#	Michael Alyn Miller <malyn@aeolon.com>	
ipcserver	600/tcp	Sun IPC server
ipcserver	600/udp	Sun IPC server
#	Bill Schiefelbein <schief@aspen.cray.com>	
syslog-conn	601/tcp	Reliable Syslog Service
syslog-conn	601/udp	Reliable Syslog Service
#	RFC 3195	
xmlrpc-beep	602/tcp	XML-RPC over BEEP
xmlrpc-beep	602/udp	XML-RPC over BEEP
#	RFC3529 <ftp://ftp.isi.edu/in-notes/rfc3529.txt> March 2003	
idxp	603/tcp	IDXP
idxp	603/udp	IDXP
#	RFC-ietf-idwg-beep-idxp-07.txt	
tunnel	604/tcp	TUNNEL
tunnel	604/udp	TUNNEL
#	RFC3620	
soap-beep	605/tcp	SOAP over BEEP
soap-beep	605/udp	SOAP over BEEP
#	RFC3288 <ftp://ftp.isi.edu/in-notes/rfc3288.txt> April 2002	
urm	606/tcp	Cray Unified Resource Manager
urm	606/udp	Cray Unified Resource Manager
nqs	607/tcp	nqs
nqs	607/udp	nqs
#	Bill Schiefelbein <schief@aspen.cray.com>	
sift-uft	608/tcp	Sender-Initiated/Unsolicited File Transfer

sift-uft		608/udp	Sender-Initiated/Unsolicited File Transfer
#	Rick Troth <troth@rice.edu>		
npmp-trap		609/tcp	npmp-trap
npmp-trap		609/udp	npmp-trap
npmp-local		610/tcp	npmp-local
npmp-local		610/udp	npmp-local
npmp-gui		611/tcp	npmp-gui
npmp-gui		611/udp	npmp-gui
#	John Barnes <jbarnes@crl.com>		
hmmp-ind		612/tcp	HMMP Indication
hmmp-ind		612/udp	HMMP Indication
hmmp-op		613/tcp	HMMP Operation
hmmp-op		613/udp	HMMP Operation
#	Andrew Sinclair <andrsin@microsoft.com>		
sshell		614/tcp	SSLshell
sshell		614/udp	SSLshell
#	Simon J. Gerraty <sjg@quick.com.au>		
sco-inetmgr		615/tcp	Internet Configuration Manager
sco-inetmgr		615/udp	Internet Configuration Manager
sco-sysmgr		616/tcp	SCO System Administration Server
sco-sysmgr		616/udp	SCO System Administration Server
sco-dtmgr		617/tcp	SCO Desktop Administration Server
sco-dtmgr		617/udp	SCO Desktop Administration Server
#	Christopher Durham <chrisdu@sco.com>		
dei-icda		618/tcp	DEI-ICDA
dei-icda		618/udp	DEI-ICDA
#	David Turner <digital@Quetico.tbaytel.net>		
compaq-evm		619/tcp	Compaq EVM
compaq-evm		619/udp	Compaq EVM
#	Jem Treadwell <Jem.Treadwell@compaq.com>		
sco-websrvrmgr		620/tcp	SCO WebServer Manager
sco-websrvrmgr		620/udp	SCO WebServer Manager
#	Christopher Durham <chrisdu@sco.com>		
escp-ip		621/tcp	ESCP
escp-ip		621/udp	ESCP
#	Lai Zit Seng <lzs@pobox.com>		
collaborator		622/tcp	Collaborator
collaborator		622/udp	Collaborator
#	Johnson Davis <johnsond@opteamasoft.com>		
asf-rmcp		623/tcp	ASF Remote Management and Control Protocol

asf-rmcp	623/udp	ASF Remote Management and Control Protocol
#	Carl First <Carl.L.First@intel.com>	
cryptoadmin	624/tcp	Crypto Admin
cryptoadmin	624/udp	Crypto Admin
#	Tony Walker <tony@cryptocard.com>	
dec_dlm	625/tcp	DEC DLM
dec_dlm	625/udp	DEC DLM
#	Rudi Martin <Rudi.Martin@edo.mts.dec.com>	
asia	626/tcp	ASIA
asia	626/udp	ASIA
#	Michael Dasenbrock <dasenbro@apple.com>	
passgo-tivoli	627/tcp	PassGo Tivoli
passgo-tivoli	627/udp	PassGo Tivoli
#	Chris Hall <chall@passgo.com>	
qmqp	628/tcp	QMQP
qmqp	628/udp	QMQP
#	Dan Bernstein <djb@cr.yp.to>	
3com-amp3	629/tcp	3Com AMP3
3com-amp3	629/udp	3Com AMP3
#	Prakash Banthia <prakash_banthia@3com.com>	
rda	630/tcp	RDA
rda	630/udp	RDA
#	John Hadjioannou <john@minster.co.uk>	
ipp	631/tcp	IPP (Internet Printing Protocol)
ipp	631/udp	IPP (Internet Printing Protocol)
#	Carl-Uno Manros <manros@cp10.es.xerox.com>	
bmpp	632/tcp	bmpp
bmpp	632/udp	bmpp
#	Troy Rollo <troy@kroll.corvu.com.au>	
servstat	633/tcp	Service Status update (Sterling Software)
servstat	633/udp	Service Status update (Sterling Software)
#	Greg Rose <Greg_Rose@sydney.sterling.com>	
ginad	634/tcp	ginad
ginad	634/udp	ginad
#	Mark Crother <mark@eis.calstate.edu>	
rlzdbase	635/tcp	RLZ DBase
rlzdbase	635/udp	RLZ DBase
#	Michael Ginn <ginn@tyxar.com>	
ldaps	636/tcp	ldap protocol over TLS/SSL (was sldap)
ldaps	636/udp	ldap protocol over TLS/SSL (was sldap)
#	Pat Richard <patr@xcert.com>	

Modern Communications Systems

```
lanserver                       637/tcp    lanserver
lanserver                       637/udp    lanserver
#                  Chris Larsson <clarsson@VNET.IBM.COM>
mcns-sec                        638/tcp    mcns-sec
mcns-sec                        638/udp    mcns-sec
#                  Kaz Ozawa <k.ozawa@cablelabs.com>
msdp                            639/tcp    MSDP
msdp                            639/udp    MSDP
#                  Dino Farinacci <dino@cisco.com>
entrust-sps                     640/tcp    entrust-sps
entrust-sps                     640/udp    entrust-sps
#                  Marek Buchler <Marek.Buchler@entrust.com>
repcmd                          641/tcp    repcmd
repcmd                          641/udp    repcmd
#                  Scott Dale <scott@Replicase.com>
esro-emsdp                      642/tcp    ESRO-EMSDP V1.3
esro-emsdp                      642/udp    ESRO-EMSDP V1.3
#                  Mohsen Banan <mohsen@neda.com>
sanity                          643/tcp    SANity
sanity                          643/udp    SANity
#                  Peter Viscarola <PeterGV@osr.com>
dwr                             644/tcp    dwr
dwr                             644/udp    dwr
#                  Bill Fenner <fenner@parc.xerox.com>
pssc                            645/tcp    PSSC
pssc                            645/udp    PSSC
#                  Egon Meier-Engelen <egon.meier-engelen@dlr.de>
ldp                             646/tcp    LDP
ldp                             646/udp    LDP
#                  Bob Thomas <rhthomas@cisco.com>
dhcp-failover                   647/tcp    DHCP Failover
dhcp-failover                   647/udp    DHCP Failover
#                  Bernard Volz <volz@ipworks.com>
rrp                             648/tcp    Registry Registrar Protocol (RRP)
rrp                             648/udp    Registry Registrar Protocol
                                           (RRP)
#                  Scott Hollenbeck <shollenb@netsol.com>
cadview-3d                      649/tcp    Cadview-3d - streaming 3d
                                           models over the internet
cadview-3d                      649/udp    Cadview-3d - streaming 3d
                                           models over the internet
#                  David Cooper <david.cooper@oracle.com>
obex                            650/tcp    OBEX
obex                            650/udp    OBEX
#                  Jeff Garbers <FJG030@email.mot.com>
ieee-mms                        651/tcp    IEEE MMS
```

```
ieee-mms                       651/udp   IEEE MMS
#                      Curtis Anderson <canderson@turbolinux.com>
hello-port                     652/tcp   HELLO_PORT
hello-port                     652/udp   HELLO_PORT
#                      Patrick Cipiere <Patrick.Cipiere@UDcast.com>
repscmd                        653/tcp   RepCmd
repscmd                        653/udp   RepCmd
#                      Scott Dale <scott@tioga.com>
aodv                           654/tcp   AODV
aodv                           654/udp   AODV
#                      Charles Perkins <cperkins@eng.sun.com>
tinc                           655/tcp   TINC
tinc                           655/udp   TINC
#                      Ivo Timmermans <itimmermans@bigfoot.com>
spmp                           656/tcp   SPMP
spmp                           656/udp   SPMP
#                      Jakob Kaivo <jkaivo@nodomainname.net>
rmc                            657/tcp   RMC
rmc                            657/udp   RMC
#                      Michael Schmidt <mmaass@us.ibm.com>
tenfold                        658/tcp   TenFold
tenfold                        658/udp   TenFold
#                      Louis Olszyk <lolszyk@10fold.com>
#                              659       Removed (2001-06-06)
mac-srvr-admin                 660/tcp   MacOS Server Admin
mac-srvr-admin                 660/udp   MacOS Server Admin
#              Forest Hill <forest@apple.com>
hap                            661/tcp   HAP
hap                            661/udp   HAP
#              Igor Plotnikov <igor@uroam.com>
pftp                           662/tcp   PFTP
pftp                           662/udp   PFTP
#              Ben Schluricke <support@pftp.de>
purenoise                      663/tcp   PureNoise
purenoise                      663/udp   PureNoise
#              Sam Osa <pristine@mailcity.com>
asf-secure-rmcp                664/tcp   ASF Secure Remote
                                         Management and Control
                                         Protocol
asf-secure-rmcp                664/udp   ASF Secure Remote
                                Management and Control Protocol
#              Carl First <Carl.L.First@intel.com>
sun-dr                         665/tcp   Sun DR
sun-dr                         665/udp   Sun DR
#              Harinder Bhasin <Harinder.Bhasin@Sun.COM>
mdqs                           666/tcp
```

mdqs	666/udp	
doom	666/tcp	doom Id Software
doom	666/udp	doom Id Software
#	<ddt@idcube.idsoftware.com>	
disclose	667/tcp	campaign contribution disclosures - SDR Technologies
disclose	667/udp	campaign contribution disclosures - SDR Technologies
#	Jim Dixon <jim@lambda.com>	
mecomm	668/tcp	MeComm
mecomm	668/udp	MeComm
meregister	669/tcp	MeRegister
meregister	669/udp	MeRegister
#	Armin Sawusch <armin@esd1.esd.de>	
vacdsm-sws	670/tcp	VACDSM-SWS
vacdsm-sws	670/udp	VACDSM-SWS
vacdsm-app	671/tcp	VACDSM-APP
vacdsm-app	671/udp	VACDSM-APP
vpps-qua	672/tcp	VPPS-QUA
vpps-qua	672/udp	VPPS-QUA
cimplex	673/tcp	CIMPLEX
cimplex	673/udp	CIMPLEX
#	Ulysses G. Smith Jr. <ugsmith@cesi.com>	
acap	674/tcp	ACAP
acap	674/udp	ACAP
#	Chris Newman <chris.newman@sun.com>	
dctp	675/tcp	DCTP
dctp	675/udp	DCTP
#	Andre Kramer <Andre.Kramer@ansa.co.uk>	
vpps-via	676/tcp	VPPS Via
vpps-via	676/udp	VPPS Via
#	Ulysses G. Smith Jr. <ugsmith@cesi.com>	
vpp	677/tcp	Virtual Presence Protocol
vpp	677/udp	Virtual Presence Protocol
#	Klaus Wolf <wolf@cobrow.com>	
ggf-ncp	678/tcp	GNU Generation Foundation NCP
ggf-ncp	678/udp	GNU Generation Foundation NCP
#	Noah Paul <noahp@altavista.net>	
mrm	679/tcp	MRM
mrm	679/udp	MRM
#	Liming Wei <lwei@cisco.com>	
entrust-aaas	680/tcp	entrust-aaas
entrust-aaas	680/udp	entrust-aaas
entrust-aams	681/tcp	entrust-aams
entrust-aams	681/udp	entrust-aams
#	Adrian Mancini <adrian.mancini@entrust.com>	

```
xfr                         682/tcp    XFR
xfr                         682/udp    XFR
#               Noah Paul <noahp@ultranet.com>
corba-iiop                  683/tcp    CORBA IIOP
corba-iiop                  683/udp    CORBA IIOP
corba-iiop-ssl              684/tcp    CORBA IIOP SSL
corba-iiop-ssl              684/udp    CORBA IIOP SSL
#               Andrew Watson <andrew@omg.org>
mdc-portmapper              685/tcp    MDC Port Mapper
mdc-portmapper              685/udp    MDC Port Mapper
#               Noah Paul <noahp@altavista.net>
hcp-wismar                  686/tcp    Hardware Control Protocol
                                       Wismar
hcp-wismar                  686/udp    Hardware Control Protocol
                                       Wismar
#               David Merchant <d.f.merchant@livjm.ac.uk>
asipregistry                687/tcp    asipregistry
asipregistry                687/udp    asipregistry
#               Erik Sea <sea@apple.com>
realm-rusd                  688/tcp    REALM-RUSD
realm-rusd                  688/udp    REALM-RUSD
#               Jerry Knight <jknight@realminfo.com>
nmap                        689/tcp    NMAP
nmap                        689/udp    NMAP
#               Peter Dennis Bartok <peter@novonyx.com>
vatp                        690/tcp    VATP
vatp                        690/udp    VATP
#               Atica Software <comercial@aticasoft.es>
msexch-routing              691/tcp    MS Exchange Routing
msexch-routing              691/udp    MS Exchange Routing
#               David Lemson <dlemson@microsoft.com>
hyperwave-isp               692/tcp    Hyperwave-ISP
hyperwave-isp               692/udp    Hyperwave-ISP
#               Gerald Mesaric <gmesaric@hyperwave.com>
connendp                    693/tcp    connendp
connendp                    693/udp    connendp
#               Ronny Bremer <rbremer@future-gate.com>
ha-cluster                  694/tcp    ha-cluster
ha-cluster                  694/udp    ha-cluster
#               Alan Robertson <alanr@unix.sh>
ieee-mms-ssl                695/tcp    IEEE-MMS-SSL
ieee-mms-ssl                695/udp    IEEE-MMS-SSL
#               Curtis Anderson <ecanderson@turbolinux.com>
rushd                       696/tcp    RUSHD
rushd                       696/udp    RUSHD
#               Greg Ercolano <erco@netcom.com>
```

uuidgen	697/tcp	UUIDGEN
uuidgen	697/udp	UUIDGEN
#	James Falkner <james.falkner@sun.com>	
olsr	698/tcp	OLSR
olsr	698/udp	OLSR
#	Thomas Clausen <thomas.clausen@inria.fr>	
accessnetwork	699/tcp	Access Network
accessnetwork	699/udp	Access Network
#	Yingchun Xu <Yingchun_Xu@3com.com>	
epp	700/tcp	Extensible Provisioning Protocol
epp	700/udp	Extensible Provisioning Protocol
#	RFC3734	
#	701-703	Unassigned
elcsd	704/tcp	errlog copy/server daemon
elcsd	704/udp	errlog copy/server daemon
agentx	705/tcp	AgentX
agentx	705/udp	AgentX
#	Bob Natale <bob.natale@appliedsnmp.com>	
silc	706/tcp	SILC
silc	706/udp	SILC
#	Pekka Riikonen <priikone@poseidon.pspt.fi>	
borland-dsj	707/tcp	Borland DSJ
borland-dsj	707/udp	Borland DSJ
#	Gerg Cole <gcole@corp.borland.com>	
#	708	Unassigned
entrust-kmsh	709/tcp	Entrust Key Management Service Handler
entrust-kmsh	709/udp	Entrust Key Management Service Handler
entrust-ash	710/tcp	Entrust Administration Service Handler
entrust-ash	710/udp	Entrust Administration Service Handler
#	Peter Whittaker <pww@entrust.com>	
cisco-tdp	711/tcp	Cisco TDP
cisco-tdp	711/udp	Cisco TDP
#	Bruce Davie <bsd@cisco.com>	
tbrpf	712/tcp	TBRPF
tbrpf	712/udp	TBRPF
#	RFC3684	
#	713-728	Unassigned
netviewdm1	729/tcp	IBM NetView DM/6000 Server/Client
netviewdm1	729/udp	IBM NetView DM/6000 Server/Client
netviewdm2	730/tcp	IBM NetView DM/6000 send/tcp

netviewdm2		730/udp	IBM NetView DM/6000
			send/tcp
netviewdm3		731/tcp	IBM NetView DM/6000
			receive/tcp
netviewdm3		731/udp	IBM NetView DM/6000
			receive/tcp
#	Philippe Binet (phbinet@vnet.IBM.COM)		
#		732-740	Unassigned
netgw		741/tcp	netGW
netgw		741/udp	netGW
#	Oliver Korfmacher (okorf@netcs.com)		
netrcs		742/tcp	Network based Rev. Cont. Sys.
netrcs		742/udp	Network based Rev. Cont. Sys.
#	Gordon C. Galligher <gorpong@ping.chi.il.us>		
#		743	Unassigned
flexlm		744/tcp	Flexible License Manager
flexlm		744/udp	Flexible License Manager
#	Matt Christiano		
#	<globes@matt@oliveb.atc.olivetti.com>		
#		745-746	Unassigned
fujitsu-dev		747/tcp	Fujitsu Device Control
fujitsu-dev		747/udp	Fujitsu Device Control
ris-cm		748/tcp	Russell Info Sci Calendar
			Manager
ris-cm		748/udp	Russell Info Sci Calendar
			Manager
kerberos-adm		749/tcp	kerberos administration
kerberos-adm		749/udp	kerberos administration
rfile		750/tcp	
loadav		750/udp	
kerberos-iv		750/udp	kerberos version iv
#	Martin Hamilton <martin@mrrl.lut.as.uk>		
pump		751/tcp	
pump		751/udp	
qrh		752/tcp	
qrh		752/udp	
rrh		753/tcp	
rrh		753/udp	
tell		754/tcp	send
tell		754/udp	send
#	Josyula R. Rao <jrrao@watson.ibm.com>		
#		755-756	Unassigned
nlogin		758/tcp	
nlogin		758/udp	
con		759/tcp	
con		759/udp	

ns	760/tcp	
ns	760/udp	
rxe	761/tcp	
rxe	761/udp	
quotad	762/tcp	
quotad	762/udp	
cycleserv	763/tcp	
cycleserv	763/udp	
omserv	764/tcp	
omserv	764/udp	
webster	765/tcp	
webster	765/udp	
#	Josyula R. Rao <jrrao@watson.ibm.com>	
#	766	Unassigned
phonebook	767/tcp	phone
phonebook	767/udp	phone
#	Josyula R. Rao <jrrao@watson.ibm.com>	
#	768	Unassigned
vid	769/tcp	
vid	769/udp	
cadlock	770/tcp	
cadlock	770/udp	
rtip	771/tcp	
rtip	771/udp	
cycleserv2	772/tcp	
cycleserv2	772/udp	
submit	773/tcp	
notify	773/udp	
rpasswd	774/tcp	
acmaint_dbd	774/udp	
entomb	775/tcp	
acmaint_transd	775/udp	
wpages	776/tcp	
wpages	776/udp	
#	Josyula R. Rao <jrrao@watson.ibm.com>	
multiling-http	777/tcp	Multiling HTTP
multiling-http	777/udp	Multiling HTTP
#	Alejandro Bonet <babel@ctv.es>	
#	778-779	Unassigned
wpgs	780/tcp	
wpgs	780/udp	
#	Josyula R. Rao <jrrao@watson.ibm.com>	
#	781-785	Unassigned
#	786	Unassigned (Removed 2002 -05-08)
#	787	Unassigned (Removed 2002

```
                                              -10-08)
#                             788-799  Unassigned
mdbs_daemon                   800/tcp
mdbs_daemon                   800/udp
device                        801/tcp
device                        801/udp
#                             802-809  Unassigned
fcp-udp                       810/tcp    FCP
fcp-udp                       810/udp    FCP Datagram
#           Paul Whittemore <paul@softarc.com>
#                             811-827  Unassigned
itm-mcell-s                   828/tcp    itm-mcell-s
itm-mcell-s                   828/udp    itm-mcell-s
#           Miles O'Neal <meo@us.itmasters.com>
pkix-3-ca-ra                  829/tcp    PKIX-3 CA/RA
pkix-3-ca-ra                  829/udp    PKIX-3 CA/RA
#           Carlisle Adams <Cadams@entrust.com>
#                             830-846  Unassigned
dhcp-failover2                847/tcp    dhcp-failover 2
dhcp-failover2                847/udp    dhcp-failover 2
#           Bernard Volz <volz@ipworks.com>
gdoi                          848/tcp    GDOI
gdoi                          848/udp    GDOI
#           RFC-ietf-msec-gdoi-07.txt
#                             849-859  Unassigned
iscsi                         860/tcp    iSCSI
iscsi                         860/udp    iSCSI
#           RFC-draft-ietf-ips-iscsi-20.txt
#                             861-872  Unassigned
rsync                         873/tcp    rsync
rsync                         873/udp    rsync
#           Andrew Tridgell <tridge@samba.anu.edu.au>
#                             874-885  Unassigned
iclcnet-locate                886/tcp    ICL coNETion locate server
iclcnet-locate                886/udp    ICL coNETion locate server
#           Bob Lyon <bl@oasis.icl.co.uk>
iclcnet_svinfo                887/tcp    ICL coNETion server info
iclcnet_svinfo                887/udp    ICL coNETion server info
#           Bob Lyon <bl@oasis.icl.co.uk>
accessbuilder                 888/tcp    AccessBuilder
accessbuilder                 888/udp    AccessBuilder
#           Steve Sweeney <Steven_Sweeney@3mail.3com.com>
# The following entry records an unassigned but widespread use
cddbp                         888/tcp    CD Database Protocol
#           Steve Scherf <steve@moonsoft.com>
#
```

#	889-899	Unassigned
omginitialrefs	900/tcp	OMG Initial Refs
omginitialrefs	900/udp	OMG Initial Refs
#	Christian Callsen <Christian.Callsen@eng.sun.com>	
smpnameres	901/tcp	SMPNAMERES
smpnameres	901/udp	SMPNAMERES
#	Leif Ekblad <leif@rdos.net>	
ideafarm-chat	902/tcp	IDEAFARM-CHAT
ideafarm-chat	902/udp	IDEAFARM-CHAT
ideafarm-catch	903/tcp	IDEAFARM-CATCH
ideafarm-catch	903/udp	IDEAFARM-CATCH
#	Wo'o Ideafarm <1@ideafarm.com>	
#	904-910	Unassigned
xact-backup	911/tcp	xact-backup
xact-backup	911/udp	xact-backup
#	Bill Carroll <billc@xactlabs.com>	
apex-mesh	912/tcp	APEX relay-relay service
apex-mesh	912/udp	APEX relay-relay service
apex-edge	913/tcp	APEX endpoint-relay service
apex-edge	913/udp	APEX endpoint-relay service
#	[RFC3340]	
#	914-988	Unassigned
ftps-data	989/tcp	ftp protocol, data, over TLS/SSL
ftps-data	989/udp	ftp protocol, data, over TLS/SSL
ftps	990/tcp	ftp protocol, control, over TLS/SSL
ftps	990/udp	ftp protocol, control, over TLS/SSL
#	Christopher Allen <ChristopherA@consensus.com>	
nas	991/tcp	Netnews Administration System
nas	991/udp	Netnews Administration System
#	Vera Heinau <heinau@fu-berlin.de>	
#	Heiko Schlichting <heiko@fu-berlin.de>	
telnets	992/tcp	telnet protocol over TLS/SSL
telnets	992/udp	telnet protocol over TLS/SSL
imaps	993/tcp	imap4 protocol over TLS/SSL
imaps	993/udp	imap4 protocol over TLS/SSL
ircs	994/tcp	irc protocol over TLS/SSL
ircs	994/udp	irc protocol over TLS/SSL
#	Christopher Allen <ChristopherA@consensus.com>	
pop3s	995/tcp	pop3 protocol over TLS/SSL (was spop3)
pop3s	995/udp	pop3 protocol over TLS/SSL (was spop3)

Modern Communications Systems

#	Gordon Mangione <gordm@microsoft.com>	
vsinet	996/tcp	vsinet
vsinet	996/udp	vsinet
#	Rob Juergens <robj@vsi.com>	
maitrd	997/tcp	
maitrd	997/udp	
busboy	998/tcp	
puparp	998/udp	
garcon	999/tcp	
applix	999/udp	Applix ac
puprouter	999/tcp	
puprouter	999/udp	
cadlock2	1000/tcp	
cadlock2	1000/udp	
#	1001-1009	Unassigned
#	1008/udp	Possibly used by Sun Solaris????
surf	1010/tcp	surf
surf	1010/udp	surf
#	Joseph Geer <jgeer@peapod.com>	
#	1011-1022	Reserved
	1023/tcp	Reserved
	1023/udp	Reserved

Source:www.iana.org accessed on 15 April 2004

Appendix D. Default ports used by some known trojan horses:

port 0 REx
port 1 (UDP) - Sockets des Troie
port 2 Death
port 5 yoyo
port 11 Skun
port 16 Skun
port 17 Skun
port 18 Skun
port 19 Skun
port 20 Amanda
port 21 ADM worm, Back Construction, Blade Runner, BlueFire, Bmail, Cattivik FTP Server, CC Invader, Dark FTP, Doly Trojan, FreddyK, Invisible FTP, KWM, MscanWorm, NerTe, NokNok, Pinochet, Ramen, Reverse Trojan, RTB 666, The Flu, WinCrash, Voyager Alpha Force
port 22 InCommand, Shaft, Skun
port 23 ADM worm, Aphex's Remote Packet Sniffer , AutoSpY, ButtMan, Fire HacKer, My Very Own trojan, Pest, RTB 666, Tiny Telnet Server - TTS, Truva Atl
port 25 Antigen, Barok, BSE, Email Password Sender , Gip, Laocoon, Magic Horse, MBT , Moscow Email trojan, Nimda, Shtirlitz, Stukach, Tapiras, WinPC
port 27 Assasin
port 28 Amanda
port 30 Agent 40421
port 31 Agent 40421, Masters Paradise, Skun
port 37 ADM worm
port 39 SubSARI
port 41 Deep Throat , Foreplay
port 44 Arctic
port 51 Fuck Lamers Backdoor
port 52 MuSka52, Skun
port 53 ADM worm, liOn, MscanWorm, MuSka52
port 54 MuSka52
port 66 AL-Bareki
port 69 BackGate Kit, Nimda, Pasana, Storm, Storm worm, Theef
port 69 (UDP) - Pasana
port 70 ADM worm
port 79 ADM worm, Firehotcker
port 80 711 trojan (Seven Eleven), AckCmd, BlueFire, Cafeini, Duddie, Executor, God Message, Intruzzo , Latinus, Lithium, MscanWorm, NerTe, Nimda, Noob, Optix Lite, Optix Pro , Power, Ramen, Remote Shell , Reverse WWW Tunnel Backdoor , RingZero, RTB 666, Scalper, Screen Cutter , Seeker, Slapper, Web Server CT , WebDownloader
port 80 (UDP) - Penrox
port 81 Asylum

Modern Communications Systems

port 101 Skun
port 102 Delf, Skun
port 103 Skun
port 105 NerTe
port 107 Skun
port 109 ADM worm
port 110 ADM worm
port 111 ADM worm, MscanWorm
port 113 ADM worm, Alicia, Cyn, DataSpy Network X, Dosh, Gibbon, Taskman
port 120 Skun
port 121 Attack Bot, God Message, JammerKillah
port 123 Net Controller
port 137 Chode, Nimda
port 137 (UDP) - Bugbear, Msinit, Opaserv, Qaz
port 138 Chode, Nimda
port 139 Chode, Fire HacKer, Msinit, Nimda, Opaserv, Qaz
port 143 ADM worm
port 146 Infector
port 146 (UDP) - Infector
port 166 NokNok
port 170 A-trojan
port 171 A-trojan
port 200 CyberSpy
port 201 One Windows Trojan
port 202 One Windows Trojan, Skun
port 211 One Windows Trojan
port 212 One Windows Trojan
port 221 Snape
port 222 NeuroticKat, Snape
port 230 Skun
port 231 Skun
port 232 Skun
port 285 Delf
port 299 One Windows Trojan
port 334 Backage
port 335 Nautical
port 370 NeuroticKat
port 400 Argentino
port 401 One Windows Trojan
port 402 One Windows Trojan
port 411 Backage
port 420 Breach
port 443 Slapper
port 445 Nimda
port 455 Fatal Connections
port 511 T0rn Rootkit

port 513 ADM worm
port 514 ADM worm
port 515 MscanWorm, Ramen
port 520 (UDP) - A UDP backdoor
port 555 711 trojan (Seven Eleven), Phase Zero, Phase-0
port 564 Oracle
port 589 Assasin
port 600 SweetHeart
port 623 RTB 666
port 635 ADM worm
port 650 Assasin
port 661 NokNok
port 666 Attack FTP, Back Construction, BLA trojan, NokNok, Reverse Trojan,
Shadow Phyre, Unicorn, yoyo
port 667 NokNok, SniperNet
port 668 Unicorn
port 669 DP trojan , SniperNet
port 680 RTB 666
port 692 GayOL
port 700 REx
port 777 Undetected
port 798 Oracle
port 808 WinHole
port 831 NeuroticKat
port 901 Net-Devil, Pest
port 902 Net-Devil, Pest
port 903 Net-Devil
port 911 Dark Shadow, Dark Shadow
port 956 Crat Pro
port 991 Snape
port 992 Snape
port 999 Deep Throat , Foreplay
port 1000 Der Späher / Der Spaeher, Direct Connection, GOTHIC Intruder ,
Theef
port 1001 Der Späher / Der Spaeher, GOTHIC Intruder , Lula, One Windows
Trojan, Theef
port 1005 Pest, Theef
port 1008 AutoSpY, li0n
port 1010 Doly Trojan
port 1011 Doly Trojan
port 1012 Doly Trojan
port 1015 Doly Trojan
port 1016 Doly Trojan
port 1020 Vampire
port 1024 Latinus, Lithium, NetSpy, Ptakks
port 1025 AcidkoR, BDDT, DataSpy Network X, Fraggle Rock , KiLo, MuSka52,

Modern Communications Systems

NetSpy, Optix Pro , Paltalk, Ptakks, Real 2000, Remote Anything, Remote Explorer Y2K, Remote Storm, RemoteNC
port 1025 (UDP) - KiLo, Optix Pro , Ptakks, Real 2000, Remote Anything, Remote Explorer Y2K, Remote Storm, Yajing
port 1026 BDDT, Dark IRC, DataSpy Network X, Delta Remote Access , Dosh, Duddie, IRC Contact, Remote Explorer 2000, RUX The Tlc.K
port 1026 (UDP) - Remote Explorer 2000
port 1027 Clandestine, DataSpy Network X, KiLo, UandMe
port 1028 DataSpy Network X, Dosh, Gibbon, KiLo, KWM, Litmus, Paltalk, SubSARI
port 1028 (UDP) - KiLo, SubSARI
port 1029 Clandestine, KWM, Litmus, SubSARI
port 1029 (UDP) - SubSARI
port 1030 Gibbon, KWM
port 1031 KWM, Little Witch, Xanadu, Xot
port 1031 (UDP) - Xot
port 1032 Akosch4, Dosh, KWM
port 1032 (UDP) - Akosch4
port 1033 Dosh, KWM, Little Witch, Net Advance
port 1034 KWM
port 1035 Dosh, KWM, RemoteNC, Truva Atl
port 1036 KWM
port 1037 Arctic , Dosh, KWM, MoSucker
port 1039 Dosh
port 1041 Dosh, RemoteNC
port 1042 BLA trojan
port 1042 (UDP) - BLA trojan
port 1043 Dosh
port 1044 Ptakks
port 1044 (UDP) - Ptakks
port 1047 RemoteNC
port 1049 Delf, The Hobbit Daemon
port 1052 Fire HacKer, Slapper, The Hobbit Daemon
port 1053 The Thief
port 1054 AckCmd, RemoteNC
port 1080 SubSeven 2.2, WinHole
port 1081 WinHole
port 1082 WinHole
port 1083 WinHole
port 1092 Hvl RAT
port 1095 Blood Fest Evolution, Hvl RAT, Remote Administration Tool - RAT
port 1097 Blood Fest Evolution, Hvl RAT, Remote Administration Tool - RAT
port 1098 Blood Fest Evolution, Hvl RAT, Remote Administration Tool - RAT
port 1099 Blood Fest Evolution, Hvl RAT, Remote Administration Tool - RAT
port 1104 (UDP) - RexxRave
port 1111 Daodan, Ultors Trojan

port 1111 (UDP) - Daodan
port 1115 Lurker, Protoss
port 1116 Lurker
port 1116 (UDP) - Lurker
port 1122 Last 2000, Singularity
port 1122 (UDP) - Last 2000, Singularity
port 1133 SweetHeart
port 1150 Orion
port 1151 Orion
port 1160 BlackRat
port 1166 CrazzyNet
port 1167 CrazzyNet
port 1170 Psyber Stream Server , Voice
port 1180 Unin68
port 1183 Cyn, SweetHeart
port 1183 (UDP) - Cyn, SweetHeart
port 1200 (UDP) - NoBackO
port 1201 (UDP) - NoBackO
port 1207 SoftWAR
port 1208 Infector
port 1212 Kaos
port 1215 Force
port 1218 Force
port 1219 Force
port 1221 Fuck Lamers Backdoor
port 1222 Fuck Lamers Backdoor
port 1234 KiLo, Ultors Trojan
port 1243 BackDoor-G, SubSeven , Tiles
port 1245 VooDoo Doll
port 1255 Scarab
port 1256 Project nEXT, RexxRave
port 1272 The Matrix
port 1313 NETrojan
port 1314 Daodan
port 1349 BO dll
port 1369 SubSeven 2.2
port 1386 Dagger
port 1415 Last 2000, Singularity
port 1433 Voyager Alpha Force
port 1441 Remote Storm
port 1492 FTP99CMP
port 1524 Trinoo
port 1560 Big Gluck, Duddie
port 1561 (UDP) - MuSka52
port 1600 Direct Connection
port 1601 Direct Connection

port 1602 Direct Connection
port 1703 Exploiter
port 1711 yoyo
port 1772 NetControle
port 1772 (UDP) - NetControle
port 1777 Scarab
port 1826 Glacier
port 1833 TCC
port 1834 TCC
port 1835 TCC
port 1836 TCC
port 1837 TCC
port 1905 Delta Remote Access
port 1911 Arctic
port 1966 Fake FTP
port 1967 For Your Eyes Only , WM FTP Server
port 1978 (UDP) - Slapper
port 1981 Bowl, Shockrave
port 1983 Q-taz
port 1984 Intruzzo , Q-taz
port 1985 Black Diver, Q-taz
port 1985 (UDP) - Black Diver
port 1986 Akosch4
port 1991 PitFall
port 1999 Back Door, SubSeven , TransScout
port 2000 A-trojan, Der Späher / Der Spaeher, Fear, Force, GOTHIC Intruder ,
Last 2000, Real 2000, Remote Explorer 2000, Remote Explorer Y2K, Senna Spy
Trojan Generator, Singularity
port 2000 (UDP) - GOTHIC Intruder , Real 2000, Remote Explorer 2000, Remote
Explorer Y2K
port 2001 Der Späher / Der Spaeher, Duddie, Glacier, Protoss, Senna Spy
Trojan Generator, Singularity, Trojan Cow
port 2001 (UDP) - Scalper
port 2002 Duddie, Senna Spy Trojan Generator, Sensive
port 2002 (UDP) - Slapper
port 2004 Duddie
port 2005 Duddie
port 2023 Ripper Pro
port 2060 Protoss
port 2080 WinHole
port 2101 SweetHeart
port 2115 Bugs
port 2130 (UDP) - Mini BackLash
port 2140 The Invasor
port 2140 (UDP) - Deep Throat , Foreplay , The Invasor
port 2149 Deep Throat

port 2150 R0xr4t
port 2156 Oracle
port 2222 SweetHeart, Way
port 2222 (UDP) - SweetHeart, Way
port 2281 Nautical
port 2283 Hvl RAT
port 2300 Storm
port 2311 Studio 54
port 2330 IRC Contact
port 2331 IRC Contact
port 2332 IRC Contact, Silent Spy
port 2333 IRC Contact
port 2334 IRC Contact, Power
port 2335 IRC Contact
port 2336 IRC Contact
port 2337 IRC Contact, The Hobbit Daemon
port 2338 IRC Contact
port 2339 IRC Contact, Voice Spy
port 2339 (UDP) - Voice Spy
port 2343 Asylum
port 2345 Doly Trojan
port 2407 yoyo
port 2418 Intruzzo
port 2555 liOn, T0rn Rootkit
port 2565 Striker trojan
port 2583 WinCrash
port 2589 Dagger
port 2600 Digital RootBeer
port 2702 Black Diver
port 2702 (UDP) - Black Diver
port 2772 SubSeven
port 2773 SubSeven , SubSeven 2.1 Gold
port 2774 SubSeven , SubSeven 2.1 Gold
port 2800 Theef
port 2929 Konik
port 2983 Breach
port 2989 (UDP) - Remote Administration Tool - RAT
port 3000 InetSpy, Remote Shut, Theef
port 3006 Clandestine
port 3024 WinCrash
port 3031 MicroSpy
port 3119 Delta Remote Access
port 3128 Reverse WWW Tunnel Backdoor , RingZero
port 3129 Masters Paradise
port 3131 SubSARI
port 3150 Deep Throat , The Invasor, The Invasor

port 3150 (UDP) - Deep Throat , Foreplay , Mini BackLash
port 3215 XHX
port 3215 (UDP) - XHX
port 3292 Xposure
port 3295 Xposure
port 3333 Daodan
port 3333 (UDP) - Daodan
port 3410 Optix Pro
port 3417 Xposure
port 3418 Xposure
port 3456 Fear, Force, Terror trojan
port 3459 Eclipse 2000, Sanctuary
port 3505 AutoSpY
port 3700 Portal of Doom
port 3721 Whirlpool
port 3723 Mantis
port 3777 PsychWard
port 3791 Total Solar Eclypse
port 3800 Total Solar Eclypse
port 3801 Total Solar Eclypse
port 3945 Delta Remote Access
port 3996 Remote Anything
port 3996 (UDP) - Remote Anything
port 3997 Remote Anything
port 3999 Remote Anything
port 4000 Remote Anything, SkyDance
port 4092 WinCrash
port 4128 RedShad
port 4128 (UDP) - RedShad
port 4156 (UDP) - Slapper
port 4201 War trojan
port 4210 Netkey
port 4211 Netkey
port 4225 Silent Spy
port 4242 Virtual Hacking Machine - VHM
port 4315 Power
port 4321 BoBo
port 4414 AL-Bareki
port 4442 Oracle
port 4444 CrackDown, Oracle, Prosiak, Swift Remote
port 4445 Oracle
port 4447 Oracle
port 4449 Oracle
port 4451 Oracle
port 4488 Event Horizon
port 4567 File Nail

port 4653 Cero
port 4666 Mneah
port 4700 Theef
port 4836 Power
port 5000 Back Door Setup, Bubbel, Ra1d, Sockets des Troie
port 5001 Back Door Setup, Sockets des Troie
port 5002 Shaft
port 5005 Aladino
port 5011 Peanut Brittle
port 5025 WM Remote KeyLogger
port 5031 Net Metropolitan
port 5032 Net Metropolitan
port 5050 R0xr4t
port 5135 Bmail
port 5150 Pizza
port 5151 Optix Lite
port 5152 Laphex
port 5155 Oracle
port 5221 NOSecure
port 5250 Pizza
port 5321 Firehotcker
port 5333 Backage
port 5350 Pizza
port 5377 Iani
port 5400 Back Construction, Blade Runner, Digital Spy
port 5401 Back Construction, Blade Runner, Digital Spy, Mneah
port 5402 Back Construction, Blade Runner, Digital Spy, Mneah
port 5418 DarkSky
port 5419 DarkSky
port 5419 (UDP) - DarkSky
port 5430 Net Advance
port 5450 Pizza
port 5503 Remote Shell
port 5534 The Flu
port 5550 Pizza
port 5555 Daodan, NoXcape
port 5555 (UDP) - Daodan
port 5556 BO Facil
port 5557 BO Facil
port 5569 Robo-Hack
port 5650 Pizza
port 5669 SpArTa
port 5679 Nautical
port 5695 Assasin
port 5696 Assasin
port 5697 Assasin

port 5742 WinCrash
port 5802 Y3K RAT
port 5873 SubSeven 2.2
port 5880 Y3K RAT
port 5882 Y3K RAT
port 5882 (UDP) - Y3K RAT
port 5888 Y3K RAT
port 5888 (UDP) - Y3K RAT
port 5889 Y3K RAT
port 5933 NOSecure
port 6000 Aladino, NetBus , The Thing
port 6006 Bad Blood
port 6267 DarkSky
port 6400 The Thing
port 6521 Oracle
port 6526 Glacier
port 6556 AutoSpY
port 6661 Weia-Meia
port 6666 AL-Bareki, KiLo, SpArTa
port 6666 (UDP) - KiLo
port 6667 Acropolis, BlackRat, Dark FTP, Dark IRC, DataSpy Network X,
Gunsan, InCommand, Kaitex, KiLo, Laocoon, Net-Devil, Reverse Trojan,
ScheduleAgent, SlackBot, SubSeven , Subseven 2.1.4 DefCon 8, Trinity, Y3K
RAT, yoyo
port 6667 (UDP) - KiLo
port 6669 Host Control, Vampire, Voyager Alpha Force
port 6670 BackWeb Server, Deep Throat , Foreplay , WinNuke eXtreame
port 6697 Force
port 6711 BackDoor-G, Duddie, KiLo, Little Witch, Netkey, Spadeace, SubSARI,
SubSeven , SweetHeart, UandMe, Way, VP Killer
port 6712 Funny trojan, KiLo, Spadeace, SubSeven
port 6713 KiLo, SubSeven
port 6714 KiLo
port 6715 KiLo
port 6718 KiLo
port 6723 Mstream
port 6766 KiLo
port 6766 (UDP) - KiLo
port 6767 KiLo, Pasana, UandMe
port 6767 (UDP) - KiLo, UandMe
port 6771 Deep Throat , Foreplay
port 6776 2000 Cracks, BackDoor-G, SubSeven , VP Killer
port 6838 (UDP) - Mstream
port 6891 Force
port 6912 Shit Heep
port 6969 2000 Cracks, BlitzNet, Dark IRC, GateCrasher, Kid Terror, Laphex,

Net Controller, SpArTa, Vagr Nocker
port 6970 GateCrasher
port 7000 Aladino, Gunsan, Remote Grab, SubSeven, SubSeven 2.1 Gold,
Theef
port 7001 Freak88, Freak2k
port 7007 Silent Spy
port 7020 Basic Hell
port 7030 Basic Hell
port 7119 Massaker
port 7215 SubSeven, SubSeven 2.1 Gold
port 7274 AutoSpY
port 7290 NOSecure
port 7291 NOSecure
port 7300 NetSpy
port 7301 NetSpy
port 7306 NetSpy
port 7307 NetSpy, Remote Process Monitor
port 7308 NetSpy, X Spy
port 7312 Yajing
port 7410 Phoenix II
port 7424 Host Control
port 7424 (UDP) - Host Control
port 7597 Qaz
port 7626 Glacier
port 7648 XHX
port 7673 Neoturk
port 7676 Neoturk
port 7677 Neoturk
port 7718 Glacier
port 7722 KiLo
port 7777 God Message
port 7788 Last 2000, Last 2000, Singularity
port 7788 (UDP) - Singularity
port 7789 Back Door Setup
port 7800 Paltalk
port 7826 Oblivion
port 7850 Paltalk
port 7878 Paltalk
port 7879 Paltalk
port 7979 Vagr Nocker
port 7983 (UDP) - Mstream
port 8011 Way
port 8012 Ptakks
port 8012 (UDP) - Ptakks
port 8080 Reverse WWW Tunnel Backdoor, RingZero, Screen Cutter
port 8090 Aphex's Remote Packet Sniffer

port 8090 (UDP) - Aphex's Remote Packet Sniffer
port 8097 Kryptonic Ghost Command Pro
port 8100 Back streets
port 8110 DLP
port 8111 DLP
port 8127 9_119, Chonker
port 8127 (UDP) - 9_119, Chonker
port 8130 9_119, Chonker, DLP
port 8131 DLP
port 8301 DLP
port 8302 DLP
port 8311 SweetHeart
port 8322 DLP
port 8329 DLP
port 8488 (UDP) - KiLo
port 8489 KiLo
port 8489 (UDP) - KiLo
port 8685 Unin68
port 8732 Kryptonic Ghost Command Pro
port 8734 AutoSpY
port 8787 Back Orifice 2000
port 8811 Fear
port 8812 FraggleRock Lite
port 8821 Alicia
port 8848 Whirlpool
port 8864 Whirlpool
port 8888 Dark IRC
port 9000 Netministrator
port 9090 Aphex's Remote Packet Sniffer
port 9117 Massaker
port 9148 Nautical
port 9301 DLP
port 9325 (UDP) - Mstream
port 9329 DLP
port 9400 InCommand
port 9401 InCommand
port 9536 Lula
port 9561 Crat Pro
port 9563 Crat Pro
port 9870 Remote Computer Control Center
port 9872 Portal of Doom
port 9873 Portal of Doom
port 9874 Portal of Doom
port 9875 Portal of Doom
port 9876 Rux
port 9877 Small Big Brother

port 9878 Small Big Brother, TransScout
port 9879 Small Big Brother
port 9919 Kryptonic Ghost Command Pro
port 9999 BlitzNet, Oracle, Spadeace
port 10000 Oracle, TCP Door, XHX
port 10000 (UDP) - XHX
port 10001 DTr, Lula
port 10002 Lula
port 10003 Lula
port 10008 li0n
port 10012 Amanda
port 10013 Amanda
port 10067 Portal of Doom
port 10067 (UDP) - Portal of Doom
port 10084 Syphillis
port 10084 (UDP) - Syphillis
port 10085 Syphillis
port 10086 Syphillis
port 10100 Control Total, GiFt trojan, Scalper
port 10100 (UDP) - Slapper
port 10167 Portal of Doom
port 10167 (UDP) - Portal of Doom
port 10498 (UDP) - Mstream
port 10520 Acid Shivers
port 10528 Host Control
port 10607 Coma
port 10666 (UDP) - Ambush
port 10887 BDDT
port 10889 BDDT
port 11000 DataRape, Senna Spy Trojan Generator
port 11011 Amanda
port 11050 Host Control
port 11051 Host Control
port 11111 Breach
port 11223 Progenic trojan, Secret Agent
port 11225 Cyn
port 11225 (UDP) - Cyn
port 11660 Back streets
port 11718 Kryptonic Ghost Command Pro
port 11831 DarkFace, DataRape, Latinus, Pest, Vagr Nocker
port 11977 Cool Remote Control
port 11978 Cool Remote Control
port 11980 Cool Remote Control
port 12000 Reverse Trojan
port 12310 PreCursor
port 12321 Protoss

Modern Communications Systems

port 12321 (UDP) - Protoss
port 12345 Ashley, BlueIce 2000, Mypic , NetBus , Pie Bill Gates, Q-taz ,
Sensive, Snape, Vagr Nocker, ValvNet , Whack Job
port 12345 (UDP) - BlueIce 2000
port 12346 NetBus
port 12348 BioNet
port 12349 BioNet, The Saint
port 12361 Whack-a-mole
port 12362 Whack-a-mole
port 12363 Whack-a-mole
port 12623 ButtMan
port 12623 (UDP) - ButtMan, DUN Control
port 12624 ButtMan, Power
port 12631 Whack Job
port 12684 Power
port 12754 Mstream
port 12904 Rocks
port 13000 Senna Spy Trojan Generator, Senna Spy Trojan Generator
port 13013 PsychWard
port 13014 PsychWard
port 13028 Back streets
port 13079 Kryptonic Ghost Command Pro
port 13370 SpArTa
port 13371 Optix Pro
port 13500 Theef
port 13753 Anal FTP
port 14194 CyberSpy
port 14285 Laocoon
port 14286 Laocoon
port 14287 Laocoon
port 14500 PC Invader
port 14501 PC Invader
port 14502 PC Invader
port 14503 PC Invader
port 15000 In Route to the Hell, R0xr4t
port 15092 Host Control
port 15104 Mstream
port 15206 KiLo
port 15207 KiLo
port 15210 (UDP) - UDP remote shell backdoor server
port 15382 SubZero
port 15432 Cyn
port 15485 KiLo
port 15486 KiLo
port 15486 (UDP) - KiLo
port 15500 In Route to the Hell

port 15512 Iani
port 15551 In Route to the Hell
port 15695 Kryptonic Ghost Command Pro
port 15845 (UDP) - KiLo
port 15852 Kryptonic Ghost Command Pro
port 16057 MoonPie
port 16484 MoSucker
port 16514 KiLo
port 16514 (UDP) - KiLo
port 16515 KiLo
port 16515 (UDP) - KiLo
port 16523 Back streets
port 16660 Stacheldraht
port 16712 KiLo
port 16761 Kryptonic Ghost Command Pro
port 16959 SubSeven , Subseven 2.1.4 DefCon 8
port 17166 Mosaic
port 17449 Kid Terror
port 17499 CrazzyNet
port 17500 CrazzyNet
port 17569 Infector
port 17593 AudioDoor
port 17777 Nephron
port 18753 (UDP) - Shaft
port 19191 BlueFire
port 19216 BackGate Kit
port 20000 Millenium, PSYcho Files, XHX
port 20001 Insect, Millenium, PSYcho Files
port 20002 AcidkoR, PSYcho Files
port 20005 MoSucker
port 20023 VP Killer
port 20034 NetBus 2.0 Pro, NetBus 2.0 Pro Hidden, Whack Job
port 20331 BLA trojan
port 20432 Shaft
port 20433 (UDP) - Shaft
port 21212 Sensive
port 21544 GirlFriend, Kid Terror
port 21554 Exploiter, FreddyK, Kid Terror, Schwindler, Sensive, Winsp00fer
port 21579 Breach
port 21957 Latinus
port 22115 Cyn
port 22222 Donald Dick, G.R.O.B., Prosiak, Ruler, RUX The Tlc.K
port 22223 RUX The Tlc.K
port 22456 Clandestine
port 22554 Schwindler
port 22783 Intruzzo

port 22784 Intruzzo
port 22785 Intruzzo
port 23000 Storm worm
port 23001 Storm worm
port 23005 NetTrash, Oxon
port 23006 NetTrash, Oxon
port 23023 Logged
port 23032 Amanda
port 23321 Konik
port 23432 Asylum
port 23456 Clandestine, Evil FTP, Vagr Nocker, Whack Job
port 23476 Donald Dick
port 23476 (UDP) - Donald Dick
port 23477 Donald Dick
port 23777 InetSpy
port 24000 Infector
port 24289 Latinus
port 25002 MOTD
port 25002 (UDP) - MOTD
port 25123 Goy'Z TroJan
port 25555 FreddyK
port 25685 MoonPie
port 25686 DarkFace, MoonPie
port 25799 FreddyK
port 25885 MOTD
port 25982 DarkFace, MoonPie
port 26274 (UDP) - Delta Source
port 26681 Voice Spy
port 27160 MoonPie
port 27184 Alvgus trojan 2000
port 27184 (UDP) - Alvgus trojan 2000
port 27373 Charge
port 27374 Bad Blood, Fake SubSeven, li0n, Ramen, Seeker, SubSeven ,
SubSeven 2.1 Gold, Subseven 2.1.4 DefCon 8, SubSeven 2.2, SubSeven Muie,
The Saint
port 27379 Optix Lite
port 27444 (UDP) - Trinoo
port 27573 SubSeven
port 27665 Trinoo
port 28218 Oracle
port 28431 Hack´a´Tack
port 28678 Exploiter
port 29104 NETrojan, NetTrojan
port 29292 BackGate Kit
port 29559 AntiLamer BackDoor , DarkFace, DataRape, Ducktoy, Latinus, Pest,
Vagr Nocker

port 29589 KiLo
port 29589 (UDP) - KiLo
port 29891 The Unexplained
port 29999 AntiLamer BackDoor
port 30000 DataRape, Infector
port 30001 Err0r32
port 30005 Litmus
port 30100 NetSphere
port 30101 NetSphere
port 30102 NetSphere
port 30103 NetSphere
port 30103 (UDP) - NetSphere
port 30133 NetSphere
port 30303 Sockets des Troie
port 30331 MuSka52
port 30464 Slapper
port 30700 Mantis
port 30947 Intruse
port 31320 Little Witch
port 31320 (UDP) - Little Witch
port 31335 Trinoo
port 31336 Butt Funnel
port 31337 ADM worm, Back Fire, Back Orifice (Lm), Back Orifice russian,
BlitzNet, BO client, BO Facil, BO2, Freak88, Freak2k, NoBackO
port 31337 (UDP) - Back Orifice, Deep BO
port 31338 Back Orifice, Butt Funnel, NetSpy (DK)
port 31338 (UDP) - Deep BO, NetSpy (DK)
port 31339 Little Witch, NetSpy (DK), NetSpy (DK)
port 31339 (UDP) - Little Witch
port 31340 Little Witch
port 31340 (UDP) - Little Witch
port 31382 Lithium
port 31415 Lithium
port 31416 Lithium
port 31416 (UDP) - Lithium
port 31557 Xanadu
port 31745 BuschTrommel
port 31785 Hack´a´Tack
port 31787 Hack´a´Tack
port 31788 Hack´a´Tack
port 31789 Hack´a´Tack
port 31789 (UDP) - Hack´a´Tack
port 31790 Hack´a´Tack
port 31791 Hack´a´Tack
port 31791 (UDP) - Hack´a´Tack
port 31792 Hack´a´Tack

Modern Communications Systems

port 31887 BDDT
port 32000 BDDT
port 32001 Donald Dick
port 32100 Peanut Brittle, Project nEXT
port 32418 Acid Battery
port 32791 Acropolis, Rocks
port 33270 Trinity
port 33333 Prosiak
port 33545 G.R.O.B.
port 33567 liOn, T0rn Rootkit
port 33568 liOn, T0rn Rootkit
port 33577 Son of PsychWard
port 33777 Son of PsychWard
port 33911 Spirit 2000, Spirit 2001
port 34312 Delf
port 34313 Delf
port 34324 Big Gluck
port 34343 Osiris
port 34444 Donald Dick
port 34555 (UDP) - Trinoo (for Windows)
port 35000 Infector
port 35555 (UDP) - Trinoo (for Windows)
port 35600 SubSARI
port 36794 Bugbear
port 37237 Mantis
port 37651 Charge
port 38741 CyberSpy
port 38742 CyberSpy
port 40071 Ducktoy
port 40308 SubSARI
port 40412 The Spy
port 40421 Agent 40421, Masters Paradise
port 40422 Masters Paradise
port 40423 Masters Paradise
port 40425 Masters Paradise
port 40426 Masters Paradise
port 41337 Storm
port 41666 Remote Boot Tool , Remote Boot Tool
port 43720 (UDP) - KiLo
port 44014 Iani
port 44014 (UDP) - Iani
port 44444 Prosiak
port 44575 Exploiter
port 44767 School Bus
port 44767 (UDP) - School Bus
port 45092 BackGate Kit

port 45454 Osiris
port 45632 Little Witch
port 45673 Acropolis, Rocks
port 46666 Taskman
port 46666 (UDP) - Taskman
port 47017 T0rn Rootkit
port 47262 (UDP) - Delta Source
port 47698 KiLo
port 47785 KiLo
port 47785 (UDP) - KiLo
port 47891 AntiLamer BackDoor
port 48004 Fraggle Rock
port 48006 Fraggle Rock
port 48512 Arctic
port 49000 Fraggle Rock
port 49683 Fenster
port 49683 (UDP) - Fenster
port 49698 (UDP) - KiLo
port 50000 SubSARI
port 50021 Optix Pro
port 50130 Enterprise
port 50505 Sockets des Troie
port 50551 R0xr4t
port 50552 R0xr4t
port 50766 Schwindler
port 50829 KiLo
port 50829 (UDP) - KiLo
port 51234 Cyn
port 51966 Cafeini
port 52365 Way
port 52901 (UDP) - Omega
port 53001 Remote Windows Shutdown - RWS
port 54283 SubSeven , SubSeven 2.1 Gold
port 54320 Back Orifice 2000
port 54321 Back Orifice 2000, School Bus , yoyo
port 55165 File Manager trojan, File Manager trojan
port 55555 Shadow Phyre
port 55665 Latinus, Pinochet
port 55666 Latinus, Pinochet
port 56565 Osiris
port 57163 BlackRat
port 57341 NetRaider
port 57785 G.R.O.B.
port 58134 Charge
port 58339 Butt Funnel
port 59211 Ducktoy

port 60000 Deep Throat , Foreplay , Sockets des Troie
port 60001 Trinity
port 60008 liOn, T0rn Rootkit
port 60068 The Thing
port 60411 Connection
port 60551 R0xr4t
port 60552 R0xr4t
port 60666 Basic Hell
port 61115 Protoss
port 61337 Nota
port 61348 Bunker-Hill
port 61440 Orion
port 61603 Bunker-Hill
port 61746 KiLo
port 61746 (UDP) - KiLo
port 61747 KiLo
port 61747 (UDP) - KiLo
port 61748 (UDP) - KiLo
port 61979 Cool Remote Control
port 62011 Ducktoy
port 63485 Bunker-Hill
port 64101 Taskman
port 65000 Devil, Sockets des Troie, Stacheldraht
port 65289 yoyo
port 65421 Alicia
port 65422 Alicia
port 65432 The Traitor (= th3tr41t0r)
port 65432 (UDP) - The Traitor (= th3tr41t0r)
port 65530 Windows Mite
port 65535 RC1 trojan

Source:www.simovits.com

Appendix E ASCII Code Set
ASCII Table (7-bit)
(ASCII = American Standard Code for Information Interchange)
(also see Related Links below)

Decimal	Octal	Hex	Binary	Value	
000	000	000	00000000	NUL	(Null char.)
001	001	001	00000001	SOH	(Start of Header)
002	002	002	00000010	STX	(Start of Text)
003	003	003	00000011	ETX	(End of Text)
004	004	004	00000100	EOT	(End of Transmission)
005	005	005	00000101	ENQ	(Enquiry)
006	006	006	00000110	ACK	(Acknowledgment)
007	007	007	00000111	BEL	(Bell)
008	010	008	00001000	BS	(Backspace)
009	011	009	00001001	HT	(Horizontal Tab)
010	012	00A	00001010	LF	(Line Feed)
011	013	00B	00001011	VT	(Vertical Tab)
012	014	00C	00001100	FF	(Form Feed)
013	015	00D	00001101	CR	(Carriage Return)
014	016	00E	00001110	SO	(Shift Out)
015	017	00F	00001111	SI	(Shift In)
016	020	010	00010000	DLE	(Data Link Escape)
017	021	011	00010001	DC1(XON)	(Device Control 1)
018	022	012	00010010	DC2	(Device Control 2)
019	023	013	00010011	DC3(XOFF)	Device Control 3)
020	024	014	00010100	DC4	(Device Control 4)
021	025	015	00010101	NAK	(Negative Acknowledgement)
022	026	016	00010110	SYN	(Synchronous Idle)
023	027	017	00010111	ETB	(End of Trans. Block)
024	030	018	00011000	CAN	(Cancel)
025	031	019	00011001	EM	(End of Medium)
026	032	01A	00011010	SUB	(Substitute)
027	033	01B	00011011	ESC	(Escape)
028	034	01C	00011100	FS	(File Separator)
029	035	01D	00011101	GS	(Group Separator)
030	036	01E	00011110	RS	(Request to Send)(Record Separator)
031	037	01F	00011111	US	(Unit Separator)
032	040	020	00100000	SP	(Space)
033	041	021	00100001	!	(exclamation mark)
034	042	022	00100010	"	(double quote)
035	043	023	00100011	#	(number sign)
036	044	024	00100100	$	(dollar sign)
037	045	025	00100101	%	(percent)
038	046	026	00100110	&	(ampersand)
039	047	027	00100111	'	(single quote)
040	050	028	00101000	(	(left/opening parenthesis)
041	051	029	00101001	)	(right/closing parenthesis)
042	052	02A	00101010	*	(asterisk)
043	053	02B	00101011	+	(plus)
044	054	02C	00101100	,	(comma)
045	055	02D	00101101	-	(minus or dash)
046	056	02E	00101110	.	(dot)
047	057	02F	00101111	/	(forward slash)
048	060	030	00110000	0	

049	061	031	00110001	1	
050	062	032	00110010	2	
051	063	033	00110011	3	
052	064	034	00110100	4	
053	065	035	00110101	5	
054	066	036	00110110	6	
055	067	037	00110111	7	
056	070	038	00111000	8	
057	071	039	00111001	9	
058	072	03A	00111010	:	(colon)
059	073	03B	00111011	;	(semi-colon)
060	074	03C	00111100	<	(less than)
061	075	03D	00111101	=	(equal sign)
062	076	03E	00111110	>	(greater than)
063	077	03F	00111111	?	(question mark)
064	100	040	01000000	@	(AT symbol)
065	101	041	01000001	A	
066	102	042	01000010	B	
067	103	043	01000011	C	
068	104	044	01000100	D	
069	105	045	01000101	E	
070	106	046	01000110	F	
071	107	047	01000111	G	
072	110	048	01001000	H	
073	111	049	01001001	I	
074	112	04A	01001010	J	
075	113	04B	01001011	K	
076	114	04C	01001100	L	
077	115	04D	01001101	M	
078	116	04E	01001110	N	
079	117	04F	01001111	O	
080	120	050	01010000	P	
081	121	051	01010001	Q	
082	122	052	01010010	R	
083	123	053	01010011	S	
084	124	054	01010100	T	
085	125	055	01010101	U	
086	126	056	01010110	V	
087	127	057	01010111	W	
088	130	058	01011000	X	
089	131	059	01011001	Y	
090	132	05A	01011010	Z	
091	133	05B	01011011	[	(left/opening bracket)
092	134	05C	01011100	\	(back slash)
093	135	05D	01011101	]	(right/closing bracket)
094	136	05E	01011110	^	(caret/cirumflex)
095	137	05F	01011111	_	(underscore)
096	140	060	01100000	`	
097	141	061	01100001	a	
098	142	062	01100010	b	
099	143	063	01100011	c	
100	144	064	01100100	d	
101	145	065	01100101	e	
102	146	066	01100110	f	
103	147	067	01100111	g	
104	150	068	01101000	h	

105	151	069	01101001	i	
106	152	06A	01101010	j	
107	153	06B	01101011	k	
108	154	06C	01101100	l	
109	155	06D	01101101	m	
110	156	06E	01101110	n	
111	157	06F	01101111	o	
112	160	070	01110000	p	
113	161	071	01110001	q	
114	162	072	01110010	r	
115	163	073	01110011	s	
116	164	074	01110100	t	
117	165	075	01110101	u	
118	166	076	01110110	v	
119	167	077	01110111	w	
120	170	078	01111000	x	
121	171	079	01111001	y	
122	172	07A	01111010	z	
123	173	07B	01111011	{	(left/opening brace)
124	174	07C	01111100			(vertical bar)
125	175	07D	01111101	}	(right/closing brace)
126	176	07E	01111110	~	(tilde)
127	177	07F	01111111	DEL	(delete)

Appendix F EBCDIC Code Set

EBCDIC Table

Reference Information
Documentation
ASCII Table
EBCDIC Table
Hexadecimal Table

Dec	Hex	Code	Dec	Hex	Code	Dec	Hex	Code	Dec	Hex	Code
0	00	NUL	32	20		64	40	space	96	60	-
1	01	SOH	33	21		65	41		97	61	/
2	02	STX	34	22		66	42		98	62	
3	03	ETX	35	23		67	43		99	63	
4	04		36	24		68	44		100	64	
5	05	HT	37	25	LF	69	45		101	65	
6	06		38	26	ETB	70	46		102	66	
7	07	DEL	39	27	ESC	71	47		103	67	
8	08		40	28		72	48		104	68	
9	09		41	29		73	49		105	69	
10	0A		42	2A		74	4A	[	106	6A	|
11	0B	VT	43	2B		75	4B	.	107	6B	,
12	0C	FF	44	2C		76	4C	<	108	6C	%
13	0D	CR	45	2D	ENQ	77	4D	(	109	6D	_
14	0E	SO	46	2E	ACK	78	4E	+	110	6E	>
15	0F	SI	47	2F	BEL	79	4F	|!	111	6F	?
16	10	DLE	48	30		80	50	&	112	70	
17	11		49	31		81	51		113	71	
18	12		50	32	SYN	82	52		114	72	
19	13		51	33		83	53		115	73	
20	14		52	34		84	54		116	74	
21	15		53	35		85	55		117	75	
22	16	BS	54	36		86	56		118	76	
23	17		55	37	EOT	87	57		119	77	
24	18	CAN	56	38		88	58		120	78	
25	19	EM	57	39		89	59		121	79	'
26	1A		58	3A		90	5A	!]	122	7A	:
27	1B		59	3B		91	5B	$	123	7B	#
28	1C	IFS	60	3C		92	5C	*	124	7C	@
29	1D	IGS	61	3D	NAK	93	5D	)	125	7D	'
30	1E	IRS	62	3E		94	5E	;	126	7E	=

31	1F	IUS	63	3F	SUB	95	5F	^	127	7F	"
Dec	Hex	Code	Dec	Hex	Code	Dec	Hex	Code	Dec	Hex	Code
128	80		160	A0		192	C0	{	224	E0	\
129	81	a	161	A1	~	193	C1	A	225	E1	
130	82	b	162	A2	s	194	C2	B	226	E2	S
131	83	c	163	A3	t	195	C3	C	227	E3	T
132	84	d	164	A4	u	196	C4	D	228	E4	U
133	85	e	165	A5	v	197	C5	E	229	E5	V
134	86	f	166	A6	w	198	C6	F	230	E6	W
135	87	g	167	A7	x	199	C7	G	231	E7	X
136	88	h	168	A8	y	200	C8	H	232	E8	Y
137	89	i	169	A9	z	201	C9	I	233	E9	Z
138	8A		170	AA		202	CA		234	EA	
139	8B		171	AB		203	CB		235	EB	
140	8C		172	AC		204	CC		236	EC	
141	8D		173	AD		205	CD		237	ED	
142	8E		174	AE		206	CE		238	EE	
143	8F		175	AF		207	CF		239	EF	
144	90		176	B0		208	D0	}	240	F0	0
145	91	j	177	B1		209	D1	J	241	F1	1
146	92	k	178	B2		210	D2	K	242	F2	2
147	93	l	179	B3		211	D3	L	243	F3	3
148	94	m	180	B4		212	D4	M	244	F4	4
149	95	n	181	B5		213	D5	N	245	F5	5
150	96	o	182	B6		214	D6	O	246	F6	6
151	97	p	183	B7		215	D7	P	247	F7	7
152	98	q	184	B8		216	D8	Q	248	F8	8
153	99	r	185	B9		217	D9	R	249	F9	9
154	9A		186	BA		218	DA		250	FA	
155	9B		187	BB		219	DB		251	FB	
156	9C		188	BC		220	DC		252	FC	
157	9D		189	BD		221	DD		253	FD	
158	9E		190	BE		222	DE		254	FE	
159	9F		191	BF		223	DF		255	FF	

Note: Values are based off of 0-255 scale. Some COBOL functions require a 1-256 scale; add 1 for offset

Source: www.legacyj.com accessed on 18 May 2004

Modern Communications Systems

Modern Communications Systems

Bibliography

[1] Barrack, Martin K., How We Communicate: The Most Vital Skill, Glenbridge Publishing, Macomb, Ill., 1988.

[2] Bloomer, John, Power Programming with RPC, O'Reilly and Associates, Inc., Sebastopol, Ca., 1992.

[3] Campan, Alan, Col. and Dearth Douglas H., Cyberwar 2.0 Myths, Mysteries, and Reality, AFCEA International Press, Fairfax, Va. ,1998.

[4] Chowdhury, Subir, The Power of Six Sigma, Dearborn Trade, Chicago, Ill., 2001.

[5] Coulouris, G.; Dollimore, J.; and Kindberg T., Distributed Systems: Concepts and Design, 2nd Edition, Addison Wesley, Harlow, England, 1994.

[6] DOD STD 5200.28-STD, Department of Defense Trusted Computer System Evaluation Criteria, December 1985.

[7] Drucker, Peter, The Profession of Management, Harvard Business Review, Harvard Press, 2004.

[8] Fink, Donald G. and Beatty, Wayne, Standard Handbook for Electrical Engineers, 13th Edition, McGraw Hill , New York, 1993.

[9] Forbes and Mahon, Faraday, Maxwell, and the Electromagnetic Field: How Two Men Revolutionized Physics, Prometheus Books, 2014.

[10] Forouzan, Behrouz A., Data Communications and Networking, McGraw-Hill, New York, 2004.

[11] Fortier, Handbook of LAN Technology, 2nd Edition, McGraw-Hill, New York, 1992.

[12] Graf, Encyclopedia of Circuits, Vol 7, 2009.

[13] Held, Gilbert, Understanding Data Communications, 3rd Edition, SAMS, 1991.

[14] Kahn, David, The Codebreakers: The Story of Secret Writing, MacMillan, New York, 1968.

[15] Kiat, Chris Leong Wai, Software Defined Radio design for An IEEE 802.11a Transceiver using Open Source Software Communications Architecture (SCA), Naval Post Graduate School Thesis, Monterey, California, 2006.

Modern Communications Systems

[16] Martin, James, Local Area Networks: Architectures and Implementations, Prentice-Hall, New Jersey, 1989.

[17] Milligan, Christine, Netware 386 User's Guide, M & T Books, Redwood City, Ca., 1990.

[18] Murray, James D., Windows NT SNMP, O'Reilly, Sebastopol, Ca. 1998.

[19] National Computer Security Center, Trusted Network Interpretation of Trusted Computer System Evaluation Criteria, NCSC-TG-005 Version 1, 31 July 1987.

[20] NIST, FIP Pub 146 – Government Open Systems Interconnection Protocol, Gaithersburg, Maryland, 1993.

[21] Panko, Raymond, Business Data Communications, Prentice Hall, 1997.

[22] Piccioni, Feynman Simplified 1A: Basics of Physics and Newton's Laws; Everyone's Guide to the Feynman Lectures, Real Science Publishing, 2014.

[23] Prasad N.S., IBM Mainframes: Architecture and Design, McGraw-Hill, New York, 1989.

[24] Sapronov, Walter, Telecommunications and the Law, Computer Sciences Press, Rockville, Maryland, 1988.

[25] Siever, Spainhour, & Patwardhan, PERL in a Nutshell: A Desktop Reference, O'Reilly, Bejiing, 1999.

[26] Spainhour, Stephen and Quercia, Valerie, Webmaster in a Nutshell; A Desktop Quick Reference, O'Reilly & Associates, Sebastopol, Ca. ,1996.

[27] Stanek, William R., XML Pocket Consultant, Microsoft Press, Redmond, Wa. 2002.

[28] Tanenbaum, Andrews S., Computer Networks, Prentice-Hall, New Jersey, 1984.

[29] Tolstoy and Silverman, Fourier Series, Dover Publications, 1976.

[30] US Army, US Army Radio Propagation Manuals, US Army, 1986 & 2009.

[31] Van Horn, Gayle, International Shortwave Broadcast Guide Winter 2014-2015, Teak Publishing, Brasstown, NC, 2014.

[32] Zells, Lois, <u>Managing Software Projects</u>, QED, Wellesley, MA.

[33] Zimmerman, Scott, <u>Building an Intranet with Windows NT 4</u>, SAMS Net, Indianapolis, Ind., 1998.

Modern Communications Systems

Websites

Air Force – www.airforce.mil
Air Force Academy Research - www.usafa.edu
AirForce Institute of Technology – www.afit.edu
ACM – www.acm.org
American University – www.american.edu
AOC – www.aoc.org
Army – www.army.mil
COMSAT – www.comsat.com
DOJ – www.doj.gov
FAA – www.faa.gov
FBI – www.fbi.gov
FCC – www.fcc.gov
George Washington University – www.gwu.edu
Google Search Engine – www.google.com
GSA – www.gsa.gov
Harvard University – www.harvard.edu
How Stuff Works – www.howstuffworks.com
Howard Community College – www.howardcc.edu
IBM – www.ibm.com
ICCP – www.iccp.org
IEEE – www.ieee.org
Jane's Defense Systems – www.janes.com
John's Hopkins University APL – www.jhu.edu/apl
McGraw-Hill books – www.mhhe.com
NASA – propagation.grc.nasa.gov
Navy – www.navy.mil
NIST – www.nist.gov
NSA – www.nsa.gov
Oracle – www.oracle.com
Pentagon – www.pentagon.mil
SANS Institute – www.sans.org
White House – www.whitehouse.gov
University of Hawaii – www.hawaii.edu
University of Maryland – www.umd.edu
US Miluitary Academy – www.usma.edu
US Naval Academy – www.usna.edu
Ziff Davis – www.zdnet.com

Modern Communications Systems

Biography

Donald Joseph Gray Chiarella lives in Elkridge-Hanover in Howard County, Maryland with is wife and 4 great children. He attends Glen Mar United Methodist in Ellicott City, Maryland. He has served as president of the United Methodist Men's at Cheltenham and Savage United Methodist and Lobbied Congress for World Peace in 1993 in this role. Don is a direct descendent of Scottish Presbyterians (Clan Gray, Stewart, Sutherland, MacGowen of Kilmarnock, Scotland), and Polish Catholics (New York, Ohio, Connecticut) as a fourth generation American since 1890. He was baptized Catholic (Scotland), American Baptist (Topeka, Kansas), and United Methodist (Maryland by Marriage). He is in the 1997-2015 editions of Marquis Who's Who in Science and Engineering. He holds an online Ph.D. in MIS from Kennedy Western University (2001), an M.S. degree in Technology Management from American University (1988) (Dean's List), and a B.A. degree in Urban Planning / IFSM from University of Maryland (1979). His second BA is in Organizational Management and Political Sciences from Ashford University (2009). He is studying EE communications and EW at Howard Community College since 2015. He is certified by George Washington University in Government Contracting. He holds the certification of CDMP from the ICCP and CSIM from ISACA. He is also a

Professional Engineering Manager (PEM) by ASEM. He is certified in Homeland

Security as a CIA - Certified Information Assurance. He has attended online MIT

Open Courseware, Naval Post Graduate School, National Defense University,

Department of Defense Computer Institute, US Air Force Academy, and St.

Mary's College of Maryland. He is a life member of the US Air Academy

Association of Graduates Associate Member, AFCEA, and US Naval Institute.

He is also in various alumni associations, and the Institute for Transportation

Engineers, IEEE, and AOC. He lobbied the U.S. Congress for World Peace in

1992 as president of his United Methodist Men's group. He helped found

SmartTech Inc. a Virtual DOD 8a Contractor Company and The L'Enfant Plaza

Chapter of the Professional Managers Association in Washington DC. He was

Vice President of the DPMA at American University. He has worked for 38 years

for various institutions in government, academia, and private industry in many IT

positions. Don has been blessed by God to have produced 25 books, numerous

government publications and professional articles, many college and university

student graduates and disciples, college curriculums, computer systems software

and hardware builds, numerous presentations, lecturing, and speaking on

various management and technical subjects for public and private consumption.

He has held several classified IT positions under federal agencies as a

contractor. He has also managed public policy development and operations

research budgets and staff during his career and most enjoyed teaching and

coaching over the years. He is CEO of a part-time IT Mgt and training company.-

Chiarella Consulting LLC with his family. His past projects are listed in his Linked website. He loves all types of sports, music, art, reading, writing, mathematics, management, theology, teaching, law, collecting coins and stamps and the sciences.

FINIS